W9-AVT-429

Chapman & Hall/CRC Mathematical Biology and Medicine Series

EXACTLY SOLVABLE MODELS OF BIOLOGICAL INVASION

CHAPMAN & HALL/CRC

Mathematical Biology and Medicine Series

Aims and scope:

This series aims to capture new developments and summarize what is known over the whole spectrum of mathematical and computational biology and medicine. It seeks to encourage the integration of mathematical, statistical and computational methods into biology by publishing a broad range of textbooks, reference works and handbooks. The titles included in the series are meant to appeal to students, researchers and professionals in the mathematical, statistical and computational sciences, fundamental biology and bioengineering, as well as interdisciplinary researchers involved in the field. The inclusion of concrete examples and applications, and programming techniques and examples, is highly encouraged.

Proposals for the series should be submitted to one of the series editors above or directly to:
CRC Press UK
23 Blades Court
Deodar Road
London SW15 2NU
UK

Chapman & Hall/CRC Mathematical Biology and Medicine Series

EXACTLY SOLVABLE MODELS OF BIOLOGICAL INVASION

SERGEI V. PETROVSKII AND BAI-LIAN LI

Boca Raton London New York Singapore

Published in 2006 by
Chapman & Hall/CRC
Taylor & Francis Group
6000 Broken Sound Parkway NW, Suite 300
Boca Raton, FL 33487-2742

Printed in the United States of America on acid-free paper
10 9 8 7 6 5 4 3 2 1

International Standard Book Number-10: 1-58488-521-1 (Hardcover)
International Standard Book Number-13: 978-1-58488-521-4 (Hardcover)
Library of Congress Card Number 2005048466

Library of Congress Cataloging-in-Publication Data

Petrovskii, Sergei V.
Exactly solvable models of biological invasion / Sergei V. Petrovskii, Bai-Lian Li.
p. cm. -- (Chapman & Hall/CRC mathematical biology and medicine series)
ISBN 1-58488-521-1 (alk. paper)
1. Biological invasions--Mathematical models. I. Li, Bai-Lian. II. Title. III. Series.

QH353.P48 2005
577'.18'05118--dc22 2005048466

Taylor & Francis Group
is the Academic Division of T&F Informa plc.

Visit the Taylor & Francis Web site at
http://www.taylorandfrancis.com

and the CRC Press Web site at
http://www.crcpress.com

To our families

Preface

Biological invasion is one of the most challenging and important issues in contemporary ecology. Patterns of species spread, rates of spread, the impact of various biological and environmental factors and other related problems have been under intensive study for a few decades. New effective tools and approaches have been developed, important work has been done and considerable progress has been made towards better understanding of this phenomenon.

Although a lot of results regarding biological invasion were obtained through field studies and analysis of field observations, recent advances could hardly be possible without extensive use of mathematics, in particular, mathematical modeling. The reason for this has its roots in the very nature of the problem. A regular study based on manipulated field experiments is very difficult due to the virtual impossibility of reproducing the environmental and initial conditions. Laboratory experiments are often not effective due to inconsistence of spatial scales. In these situations, mathematical modeling takes, to some extent, the role that is normally played by experimental study in other natural sciences.

It should also be mentioned that the issue of biological invasion has been an inspiration for a few generations of mathematicians. Starting from classical works by Fisher (1937) and Kolmogorov et al. (1937), this subject has been fascinating ever since and eventually became one of the cornerstones for the contemporary nonlinear science.

However, the whole understanding of the meaning and implication of mathematics in scientific studies has been changing recently. Due to tremendous progress in computational technologies in the last two decades, many theoreticians working in various fields tended to consider computer experiment as their main research tool. The whole concept of mathematical modeling was reduced to numerical simulations and the role of rigorous mathematics has often been underestimated. It must be noted that, although computer experiment is a useful and powerful approach, it has a few serious drawbacks and its actual capability is not as almighty as it is sometimes viewed. One of the drawbacks is the impact of numerical/approximation error which is usually very difficult to estimate and which, in applied numerical simulations, is practically never addressed so that its actual impact on simulation results in most cases remains obscure. Meanwhile, there are examples when the approximation error changes results essentially, especially when the problem under study is nonlinear. Another drawback is that, since even a very detailed simulation study uses a finite number of parameter values, computer experiment alone is not capable of providing full and complete information about the solution dependence on the parameters. The information obtained numerically is more meager the more parameters the problem depends on.

These difficulties never arise when the problem can be solved analytically. It should be mentioned that, when a nonlinear problem is described by partial differential equations (PDEs), the exact solution is usually a special solution, i.e., the solution possessing certain symmetry and/or obtained for special initial and boundary conditions. (Probably the most well-known example of a special solution with an immediate application to biological invasion is a traveling population front.) Remarkably, however, the meaning of such solutions is not exhausted by a few special situations and appears much more general due to initial conditions convergence which is often considered as a manifestation of a universal principle of scaling.

Besides providing an immediate and complete description of system dynamics for a relevant class of initial conditions (apart from an early stage of dynamics when the initial details can be significant), exact solutions can also infer information about a larger family of related problems via application of the comparison principle. Another useful application of exact solutions which should not be forgotten is that they serve as a convenient tool for testing complicated numerical algorithms and codes used in more specialized studies.

In this book, our main attention is given to the models based on nonlinear PDEs, especially PDEs of diffusion-reaction type. One of the reasons why the implication of exactly solvable nonlinear models in ecology has been underestimated is the widely spread opinion that such models are exceptionally rare. This is not true: indeed, they are relatively rare but not at all exceptional. Along with several ad hoc methods, there are some regular mathematical approaches that can be applied to obtain exact solutions of nonlinear PDEs. One of the goals of this book is to provide a unified description of these methods in order to bring them into use by a wider community of theoreticians working on species dispersal and biological invasion. Apart from the methods and relevant examples, we also give a review of exactly solvable models (now scattered over periodic literature) that can be useful for studying biological invasion and species spread.

Although this book is mostly concerned with mathematical aspects of invasive species spread, we never forget that there is nature standing behind. Based on exactly solvable models, we make a new insight into a few issues of significant current interest such as the impact of the Allee effect, the impact of predation, the interplay between different modes of species dispersal, etc. Thus, we expect that our book will be interesting and useful not only to applied mathematicians but also to those biologists who are not scared by differential equations. In order to make it intelligible to researchers from different fields as well as to postgraduate students, we provide as many calculation details as possible. Also, an Appendix is added giving a brief review of some background mathematics. The chapters that contain a somewhat more complicated mathematics and can be skipped for the first reading are marked with asterisk.

The book is organized as follows. In Chapter 1, we begin with providing arguments to show why exactly solvable models are important. In particular, we give a few examples when a straightforward numerical study of relevant ecological problems can be misleading. We then give a brief review of basic facts from population dynamics that should be taken into account when formulating mathematical models of biological invasion. In Chapter 2, we give an overview of several modeling approaches that have been developed and successfully applied over the last two decades. We briefly discuss the biological background behind different approaches in order to better understand their applicability and try to reveal mathematical relation between different models. In Chapter 3, we describe a few methods that are used to construct exact solutions of nonlinear diffusion-reaction equations and consider a few instructive examples. In Chapters 4 to 6, we consider several more specific ecological problems described by exactly solvable models. In Chapter 7, we show how the predictions of exactly solvable models can sometimes be extended beyond their presumptions and what alternative analytical approaches are possible in non-integrable cases. In Chapter 8, we demonstrate the usefulness of exactly solvable models by means of applying them to a few particular cases of biological invasions. Finally, the last chapter gives some basic facts about the mathematical tools used throughout the book.

This preface would not be complete if we did not mention those who have greatly helped us in our work. We think that this is a wonderful opportunity to express our sincere gratitude to the people who were our teachers at different stages of our life and whose influence on our research cannot be overestimated. We are especially thankful to Grigorii Barenblatt, Simon Levin, Nanako Shigesada and Wally Wu who gave us timeless ideas about the true standards of scientific study. We are very thankful to our colleagues and friends who throughout all the years have collaborated with us, including Horst Malchow, Rod Blackshaw, Ezio Venturino, Sasha Medvinsky, Zhen-Shan Lin, Andrew Morozov, Frank Hilker and Igor Nazarov. Many thanks are also due to Britta Daudert for her careful reading of the complete draft. We are very grateful for the valuable support and patience of our families; they understood and shared our joy in writing this book. Last but not least, we are thankful to Sunil Nair, Publisher of Mathematics and Statistics, CRC Press/Chapman & Hall, for inviting us to write this book and for his patience and encouragement.

S.V.P. & B.L.L.
Moscow and Riverside

Contents

List of Figures

Chapter 1

Introduction

1.1 Why exactly solvable models are important

Mathematical modeling has a very long history. From the golden age of ancient Greece, through controversial medieval times, and up to the great discoveries of the twentieth century, people have been fascinated by nature and tried to conceive the ways it works by means of describing its complexity with simple cogitable relations.

The progress in understanding nature has always been tightly related to progresses in mathematics. Probably the most famous attempt to build a mathematical model in the whole history of science, and also the one that took the longest time, has been that related to the ancient problem about Achilles and the turtle: whether a faster moving body can catch up with a slower one, especially if we consider their motion in small time intervals. Nowadays, the question may seem trivial but it had appeared impossible to answer it properly until the mathematical theory of infinitesimal values was developed by Isaac Newton and Gottfried von Leibniz.

The industrial revolution and considerable progress in engineering brought to life a variety of much more complicated problems. Correspondingly, theories and models were becoming more and more elaborated and that greatly stimulated further advances in mathematics. In particular, since most of the dynamical models used to be based on differential relations, advances in the theory of differential equations were substantial and it became a mature science with a comprehensive insight into its subject and a well-developed array of powerful analytical tools.

Remarkably, a growing consciousness of the complexity of nature and a general tendency for modeling approaches to become more and more complicated did not eradicate the original craving for simplicity. Perhaps the highest point achieved in this direction is the law of radioactive decay. Radioactivity is an extremely complicated phenomenon. Its full comprehension requires quantum theory and nuclear physics; yet as far as we are concerned with the "mean-field" behavior, it can be very well described by a onefold differential equation.

In the middle of the last century, a general progress in science and education resulted in a scientific revolution: apart from its traditional applications in engineering, physics and chemistry, mathematics started penetrating into other fields such as biology, ecology and social sciences. In particular, from a rather descriptive field of knowledge several decades ago, ecology has now grown to a qualitative science where mathematics is used widely and successfully.

This process has been greatly enhanced by the emergence and spread of computers. Tremendous recent progress in computer science and information technologies has equipped ecologists with a new very powerful and convenient research tool. Although an adequate ecological study must always be based on the reality check such as field observations and experiments, results of computer experiments have often proved to be very helpful in refining the goals and approaches and in providing a deeper insight into the problems under study. Several new modeling techniques have appeared that are essentially based on computer simulations, e.g., coupled lattice maps, individual-based modeling, etc. So strong is the belief in the power of computers that alternative analytical approaches are even regarded as outdated sometimes. There is a growing number of researchers working in theoretical ecology and ecological modeling who are simply not aware of existing research tools based on rigorous mathematics.

However, the enthusiasm brought forward by the new research opportunities associated with modern computers can be rather misleading sometimes. One essential drawback of the approach based on numerical simulations is that, since every particular simulation run is done for particular parameter values, computational experiments are not able to give a full and comprehensive account about the model or solution properties. Let us consider a situation when a relevant solution behavior takes place in a narrow parameter range. In case we do not have any a priori information regarding where this range is possibly situated in the parameter space of the model, the chances to find it eventually in the course of numerical experiments may be pretty low, especially when the model contains several parameters. Moreover, a regular study is thus changed, in some sense, into a "try and guess" approach and it can hardly be acceptable as an appropriate way to conduct scientific research.

It should be mentioned that the situation when a given property may take place in a narrow parameter range does not at all mean that this property is not interesting or irrelevant. Ecosystems are complex self-organized units and it is a big and largely open question why the parameter values are what they actually are.

Another source of trouble is rooted in the approximation error. Usually, it is small and can be neglected. However, the situation may be different in a multi-scale problem where both "large" and "small" variables are important. In such cases, the magnitude of the computational error is mainly determined by calculation of larger values and the resulting error is sufficiently large to blur the computational results for the smaller variables. One possible example of this situation is immediately found in the patterns of species invasion. It is well known that the rate of spread is in some cases related to the large-distance asymptotics where the population density is vanishingly small, and it is substantially different for different types of solution decay (cf. Kot et al., 1996; Sherratt and Marchant, 1996). The computational error is likely to deform the solution behavior near the leading edge and thus to modify the rate and the whole pattern of spread.

As a matter of fact, the impact of the approximation error can be even more significant, not only distorting the system properties but changing them essentially. In order to gain an insight into this possible pitfall, let us consider a finite-difference approximation of the diffusion equation in one spatial dimension:

$$\frac{\partial u}{\partial t} = \frac{\partial}{\partial x}\left(D\frac{\partial u}{\partial x}\right) \tag{1.1}$$

where x is space, t is time and u is the density of the diffusing substance. We consider Eq. (1.1) in a domain $0 < x < L$ and for $t > 0$.

Choosing a homogeneous numerical grid in space and time, $x_{i+1} = x_i + h,\ i = 0, \dots, N,\ x_0 = 0,\ x_{N+1} = L$ and $t_{k+1} = t_k + \tau,\ k = 0, 1, \dots,\ t_0 = 0$, using then a simple explicit scheme so that

$$\frac{\partial u}{\partial x} = \frac{u_{i+1,k} - u_{ik}}{h}, \qquad \frac{\partial u}{\partial t} = \frac{u_{i,k+1} - u_{ik}}{\tau} \tag{1.2}$$

where $u_{ik} = u(x_i, t_k)$, and assuming for simplicity that D is a constant coefficient, Eq. (1.1) takes the form

$$u_{i,k+1} = u_{ik} + \left(\frac{D\tau}{h^2}\right)(u_{i+1,k} + u_{i-1,k} - 2u_{ik}) \tag{1.3}$$

which makes it possible to calculate the density on the next time-layer.

Now, to what extent is Eq. (1.3) equivalent to the original Eq. (1.1)? In order to address this question, let us expand all variables contained in (1.3) into the Taylor series in vicinity of (x_i, t_k). Then, after a little algebra, we obtain:

$$\left(\frac{\partial u}{\partial t}\right)_{ik} + \tau\left[\frac{1}{2}\left(\frac{\partial^2 u}{\partial t^2}\right)_{ik} + \frac{1}{6}\left(\frac{\partial^3 u}{\partial t^3}\right)_{ik}\tau + \dots\right] \tag{1.4}$$

$$= D\left(\frac{\partial^2 u}{\partial x^2}\right)_{ik} + Dh^2\left[\frac{1}{12}\left(\frac{\partial^4 u}{\partial x^4}\right)_{ik} + \frac{1}{360}\left(\frac{\partial^6 u}{\partial x^6}\right)_{ik} h^2 + \dots\right].$$

The expressions in square brackets include higher derivatives that are absent in the original equation. Obviously, in the limiting case $h \to 0,\ \tau \to 0$, Eq. (1.4) approaches (1.1); however, for any fixed value of h and τ their properties can be essentially different. That happens in the case of constant diffusivity and, for D depending on space, time and/or density, discrepancy between the two equations can be even greater.

As an instructive example, now we consider the following problem (Barenblatt et al., 1993). It is well known that the dynamics of marine ecosystems is to a large extent controlled by water temperature and thus the related issues of heat and mass transfer in the upper productive ocean layer are of considerable interest. In a sea or ocean, the transport processes are mainly driven by turbulence, and the coefficient of turbulent exchange appears to be a function of the temperature gradient. In order to describe vertical temperature distribution, the following nonlinear diffusion equation was suggested by several authors:

$$\frac{\partial T}{\partial t} = \frac{\partial}{\partial x}\left[\phi\left(\frac{\partial T}{\partial x}\right)\right] \tag{1.5}$$

where T is water temperature and ϕ is the absolute value of the temperature flux. Analysis of experimental data indicates that ϕ depends on the temperature gradient in a nonmonotonous way, tending to zero for small and large $\partial T/\partial x$ and reaching its maximum value for an intermediate value.

Equation (1.5) was studied numerically by several authors, e.g., see Posmentier (1977) and Djumagazieva (1983), and a realistic step-like temperature distribution was obtained. However, somewhat later it was shown analytically by Höllig (1983) that Eq. (1.5) is ill-posed and the corresponding boundary problem has an infinite number of solutions – contrary to the unique solution actually obtained in numerical simulations.

The paradox was finally resolved by Barenblatt et al. (1993) who showed that Eq. (1.5) is inadequate because it neglects the time of temperature flux response to the transient temperature gradient. This time is indeed small but not at all zero. They showed that a physically correct model of vertical turbulent heat and mass transfer in a turbulent flow should contain one more term accounting for the temperature flux relaxation:

$$\frac{\partial T}{\partial t} = \frac{\partial}{\partial x}\left[\phi\left(\frac{\partial T}{\partial x}\right)\right] + \tau_0 \frac{\partial^2}{\partial t \partial x}\left[\psi\left(\frac{\partial T}{\partial x}\right)\right] \tag{1.6}$$

where τ_0 is a characteristic relaxation time and ψ is a certain function. It was then proved rigorously that the presence of the third-order derivative makes the problem well-posed and Eq. (1.6) has a unique solution. Numerical experiments confirmed that the solution properties are in a good agreement with those of the temperature field observed in a real ocean.

Now, comparison between Eqs. (1.5) and (1.6) readily leads to understanding how the "bad" equation (1.5) could be successfully solved numerically. A finite-difference approximation of the diffusion equation includes higher derivatives, cf. (1.4), and de facto regularization of the ill-posed problem was, in fact, a purely numerical artifact.

Apparently, most of the difficulties described above can hardly ever arise when a problem is solved analytically. In the rest of this book, we will show how exact solutions can be obtained for nonlinear diffusion-reaction equations used in mathematical ecology, reveal their main properties and show how they contribute to study of species spread and biological invasion.

1.2 Intra- and inter-species interactions and local population dynamics

We begin with a brief introduction into the models of local population dynamics. Only as many details will be given as it is necessary for understanding the rest of the book. For those who may be interested in getting more information, there is extensive literature concerned with this and related issues, e.g., see Kot (2001).

Populations of biological species rarely remain invariable. Typically, they change with time due to birth and death of individuals they consist of. One of the basic assumptions of theoretical population dynamics, which is in good agreement with results of many laboratory experiments and field observations, is that the rate of population growth is a function of the population density u:

$$\frac{du}{dt} = uf(u) = F(u) \tag{1.7}$$

where $f(u)$ is the per capita growth rate that can also depend on the population density. Here the right-hand side is assumed to take into account both population multiplication and natural mortality.

The properties of particular models depend on what assumptions are made regarding function f (or, rather, what factors affecting the dynamics of a given population are taken into account). The simplest approach assumes that density-dependence is absent, i.e., $f(u) = \alpha = \text{const}$; see Fig. 1.1. That immediately results in unbounded exponential growth of the population. Although it is apparently nonrealistic, if considered for any t, the growth close to an exponential can be sometimes observed for real populations but only during a relatively short time. As the population density increases, intraspecific competition is gradually becoming more and more important which results in a decrease of the per capita growth rate. For sufficiently large population densities, the growth rate will likely stop, or even may become negative when mortality prevails over population multiplication. It means that the growth rate $F(u)$ should possess the following properties:

$$F(0) = F(K) = 0, \tag{1.8}$$

$$F(u) > 0 \quad \text{for} \quad 0 < u < K, \qquad F(u) < 0 \quad \text{for} \quad u > K, \tag{1.9}$$

$$F'(0) = \alpha > 0, \qquad F'(u) < \alpha \quad \text{for} \quad u > 0 \tag{1.10}$$

where prime denotes the derivative with respect to u. Here parameter α has the meaning of the per capita population growth at small population density and parameter K is the population carrying capacity so that for $u = K$ the population growth stops; see Fig. 1.2. Condition (1.10) thus ensures that the per capita growth rate f reaches its maximum at $u = 0$.

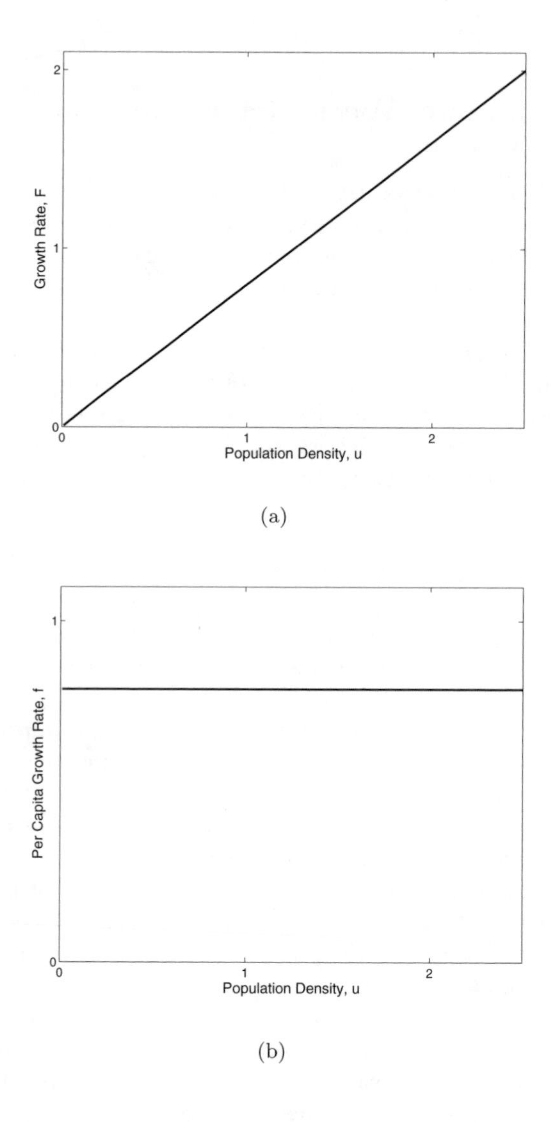

FIGURE 1.1: Population growth rate $F(u)$ and per capita growth rate $f(u)$, (a) and (b) respectively, in the absence of density-dependence.

The above relations do not assume any specific functional parameterization. Only very few results can be obtained in a general case and, to make a more detailed study, we have to choose a particular form of F (or f). Perhaps the most famous, and also the simplest model of population dynamics taking into account the impact of the intraspecific competition, is the model of logistic growth. This model assumes that the per capita growth rate decreases linearly

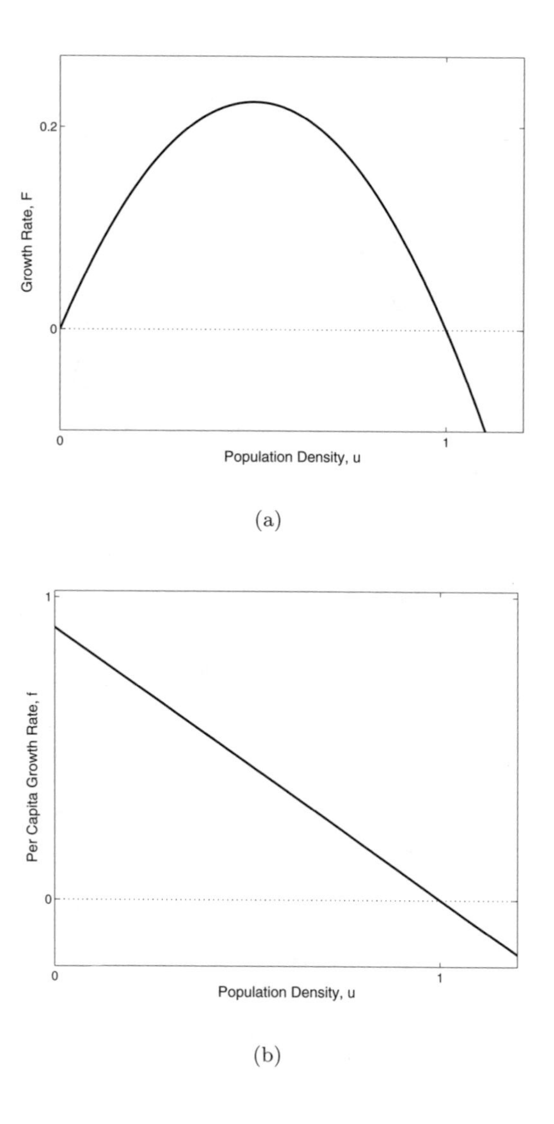

FIGURE 1.2: A sketch of the population growth rate $F(u)$ and per capita growth rate $f(u)$, (a) and (b) respectively, in the simplest case of density-dependence (logistic growth) with $K = 1$.

with population density, i.e., $f(u) = \alpha(1 - u/K)$. Equation (1.7) now reads as follows:

$$\frac{du}{dt} = \alpha u \left(1 - \frac{u}{K}\right). \tag{1.11}$$

Correspondingly, function F described by conditions (1.8–1.10) is sometimes

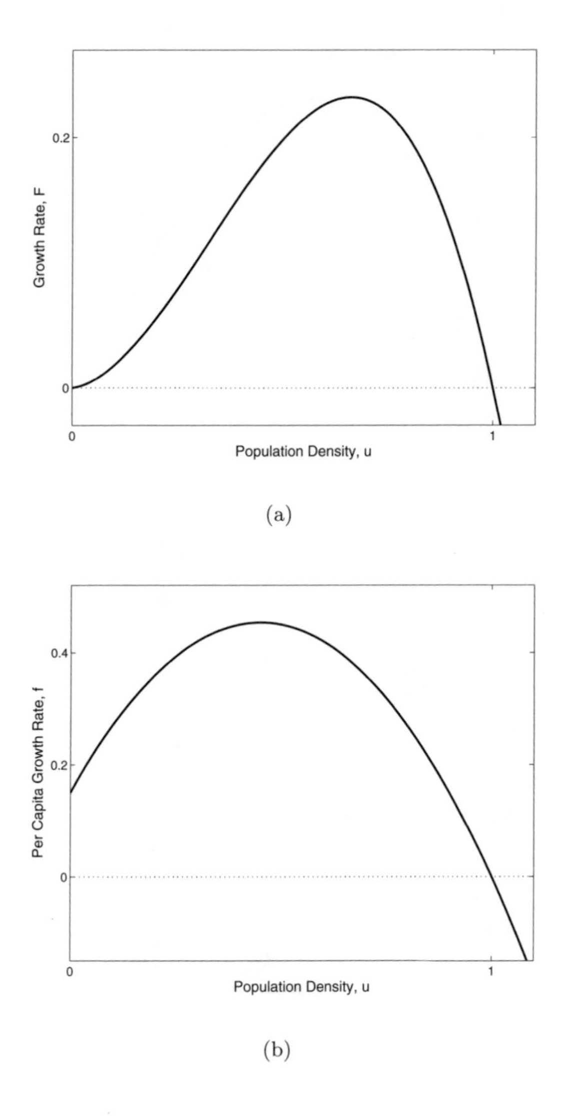

FIGURE 1.3: A sketch of the population growth rate $F(u)$ and per capita growth rate $f(u)$, (a) and (b) respectively, in the case of weak Allee effect.

referred to as a "generalized logistic growth."

Note that conditions (1.8–1.10) imply that population density u can be scaled to the population carrying capacity, $u \to \tilde{u} = u/K$. For the new variable $\tilde{u}$, Eq. (1.11) takes a somewhat more convenient form:

$$\frac{d\tilde{u}}{dt} = \alpha\tilde{u}(1 - \tilde{u}) \ . \tag{1.12}$$

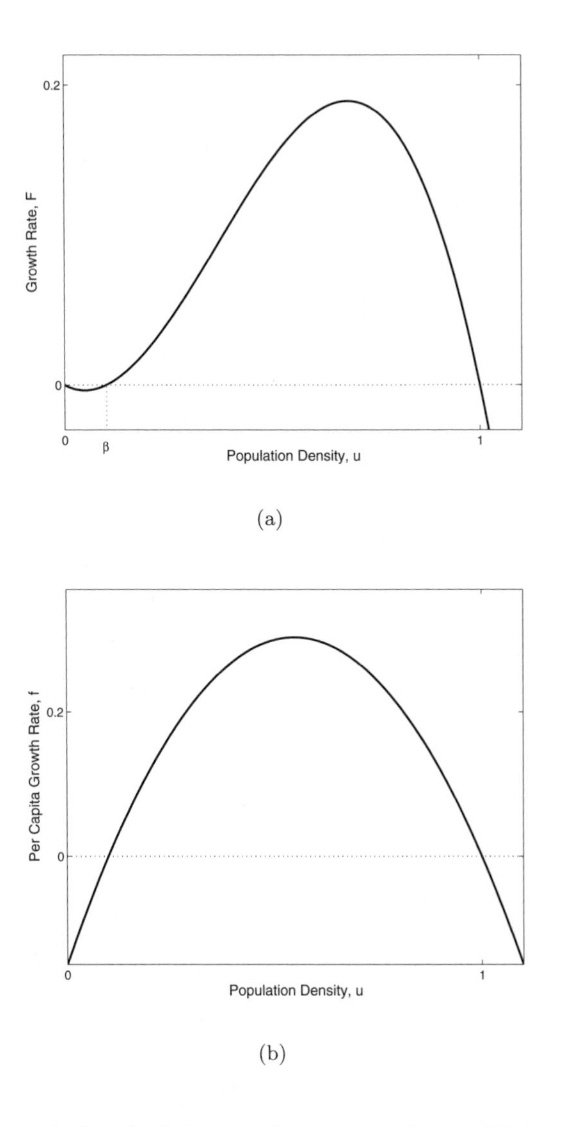

FIGURE 1.4: A sketch of the population growth rate $F(u)$ and per capita growth rate $f(u)$, (a) and (b) respectively, in the case of strong Allee effect.

According to the model of logistic growth, the per capita growth rate should decrease monotonously with the population density. There is, however, another type of density-dependence when the maximum per capita growth rate is reached for a certain intermediate density; see Fig.1.3. This shift is called the Allee effect and the corresponding population dynamics is called the Allee dynamics (Allee, 1983; Dennis, 1989; Lewis and Kareiva, 1993). From a geometrical aspect, due to the Allee effect the growth function $F(u)$ becomes concave in a vicinity of $u = 0$.

In theoretical studies, it appears convenient to distinguish between the "weak" Allee effect, cf. Fig. 1.3, and the "strong" Allee effect, see Fig. 1.4, when the population growth becomes negative for small population density. A rather general description of the growth function $F(u)$ in the case of the strong Allee effect is as follows:

$$F(u) < 0 \quad \text{for} \quad 0 < u < u_A \quad \text{and} \quad u > K, \tag{1.13}$$

$$F(u) > 0 \quad \text{for} \quad u_A < u < K. \tag{1.14}$$

Here parameter u_A corresponds to a certain threshold population density. In more focused studies, the growth rate is often described by a cubic polynomial:

$$F(u) = \omega u(u - u_A)(K - u). \tag{1.15}$$

Note that this parameterization actually makes it possible to include also the weak Allee effect. Namely, assuming that u_A can be negative (when it, of course, does not have the meaning of population density any more), the Allee effect is strong for $0 < u_A < 1$ and weak for $-1 < u_A \leq 0$. It is readily seen that for $u_A \leq -1$ the Allee effect is absent.

As well as in the case of logistic growth, the population density can be conveniently scaled to the carrying capacity; Eq. (1.7) then turns to

$$\frac{d\tilde{u}}{dt} = \gamma\tilde{u}(\tilde{u} - \beta)(1 - \tilde{u}) \tag{1.16}$$

where $\tilde{u} = u/K$, $\beta = u_A/K$ and $\gamma = K^2\omega$.

The above equations arise when we are concerned with the dynamics of a particular species. In ecological reality, however, the species interact with each other and none of them can be considered separately. Correspondingly, compared to the single-species models, the next level of model complexity appears when we consider the dynamics of a two-species community:

$$\frac{du_1}{dt} = F(u_1) + \kappa_{12}R_1(u_1, u_2)\ , \tag{1.17}$$

$$\frac{du_2}{dt} = G(u_2) + \kappa_{21}R_2(u_1, u_2) \tag{1.18}$$

where u_1 and u_2 are the species densities, $F(u_1)$ and $G(u_2)$ describe the multiplication and mortality of species 1 and 2, respectively, in the absence of other species, positively defined functions R_1 and R_2 describe the inter-species interaction and κ_{12}, κ_{21} are coefficients.

System (1.17–1.18) is rather general and, as such, it accounts for a variety of ecological situations. In particular, depending on the sign of coefficients κ_{12} and κ_{21}, it describes inter-species interactions of different type. While $\kappa_{12} = \kappa_{21} = -1$ corresponds to species competition, for $\kappa_{12} = \kappa_{21} = +1$ it describes a mutualistic community. Note that, in a general case, functions R_1 and R_2 must not necessarily coincide with each other, cf. "asymmetric competition." The case of κ_{12} and κ_{21} being of different sign corresponds to a predator-prey system; considering species 1 as prey and species 2 as predator and assuming that the predator cannot survive in the absence of prey, Eqs. (1.17–1.18) take a more specific form:

$$\frac{du_1}{dt} = f(u_1)u_1 - r(u_1)u_1u_2 \ , \tag{1.19}$$

$$\frac{du_2}{dt} = \kappa_{21}r(u_1)u_1u_2 - g(u_2)u_2 \tag{1.20}$$

where $0 < \kappa_{21} < 1$ now has the meaning of food utilization coefficient and function g gives the per capita predator mortality. Here the form of $r(u_1)$ depends on the type of predator response to prey, e.g., Holling II or Holling III.

Obviously, in general, two-species models contain more information about the population dynamics than single-species ones. It should be mentioned, however, that single-species models may also account for the impact of other species through the choice of corresponding parameter values. For instance, the impact of predation can be, to some extent, taken into account by means of increased mortality rate, the impact of competition can be described by means of choosing lower multiplication rate, etc.

1.3 Basic mechanisms of species transport

In real ecological populations, the population density is normally varying not only in time but also in space. There are many reasons for that. Heterogeneity of a population spatial distribution can arise due to heterogeneity in controlling factors, such as the growth rate and/or mortality, which results in population density changing with different rates at different locations. Similarly, it can be a result of spatially different initial conditions. It can arise also due to the impact of spatially heterogeneous stochastic factors of either environmental or demographic origin.

The above reasons are more related to the temporal dynamics of the populations and, in fact, do not account for space explicitly. Another source of the species' spatial heterogeneity is found in the populations re-distribution in space due to the transport of individuals. As well as growth, multiplication and mortality, it is a very general phenomenon with profound ecological implication and the dynamics of any population is affected significantly by transport processes. This remains true even when individuals of given species are actually immobile, e.g., for plant species. In the latter case, spatial re-distribution of the population takes place by means of seeds spreading.

Apparently, the scenarios and mechanisms of species transport as well as their implication can differ greatly from species to species and from case to case. Among other things, these differences define the time-scale where the impact of species transport on the population distribution can become important. For species with a pronounced ability for self-motion, like mammals, birds and some insects, the population spatial distribution can change remarkably during a relatively short interval falling in between two successive generations, e.g., one year. On the contrary, for a tree species, a characteristic re-distribution time should cover a few generations and can be as long as decades or even centuries.

It should be mentioned that the impact of transport processes on the population distribution is ambilateral. Species transport can result in spatial heterogeneity but it can also be caused by it. In particular, in the case of random motion of individuals, a difference between the population density in neighboring sites generates a population density flux directed toward the site with lower density. The impact of behavioral traits specific to a given species, such as an optimal foraging strategy, social or learning behavior, when individual motion can hardly be regarded as random, is likely to modify the flux intensity but, as far as both sites are equal in their "quality," will not eliminate it. In this and similar cases, the transport processes tend to decrease spatial heterogeneity rather than generate it.

In general, the mechanisms of species transport can be classified into a few types. First, there is transport caused by environmental factors such as wind for air-borne species or current for water-borne species. Considering by way of example a strongly idealized case when all individuals of a given population move with a constant speed $\mathbf{A}_0$, the population re-distribution in space is described by the following equation:

$$\frac{\partial u}{\partial t} + A_0 \frac{\partial u}{\partial x} = 0 \tag{1.21}$$

where $A_0 = |\mathbf{A}_0|$ and axis x is chosen in the direction of species motion. Equation (1.21) has a solution describing a traveling wave propagating along axis x with speed A_0, i.e., $u(x,t) = \phi(\xi)$ where $\xi = x - A_0 t$ and the wave profile ϕ is determined by the initial conditions; see Fig. 1.5. Note that in this section we do not take into account population growth and mortality. A full consideration of the corresponding population dynamics will be done in the next chapter. From the point of time-scale, Eq. (1.21) may correspond to species transport in the between-generation time.

The second type is species transport due to self-motion of individuals. Individual motion is usually intricate and is affected by a variety of environmental and biological factors (Turchin, 1998). However, from the modeling standpoint, the complexity of its description depends on how much information we actually want to gain. While an individual path can indeed be very complicated, an averaged "displacement" of a given individual may behave in a much simpler way. A conceptual assumption that is usually made is that

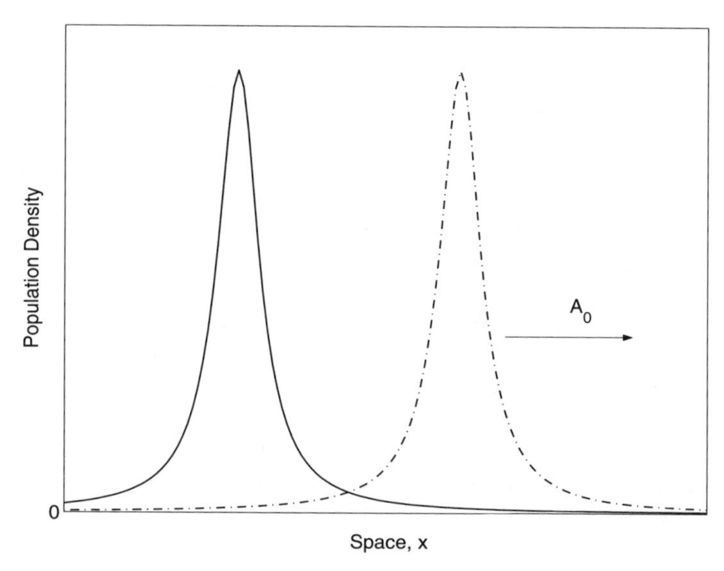

FIGURE 1.5: A sketch of a traveling population wave propagating with constant speed A_0. The solid curve shows the population density for $t = 0$; the dashed-and-dotted curve shows the population density for a certain $t = t_1 > 0$. Note that the hump moves to the right without changing its shape.

it can be regarded as random (Skellam, 1951; Okubo, 1986); the population dynamics is then described by the diffusion equation, cf. (1.1). (In this context, the population spatial re-distribution due to random individual motion is often referred to as "biodiffusion.") Note that applicability of the diffusion equation is also enhanced by the "vertical integration," i.e., by the transition from individual to population level, due to additional averaging over different individual behavioral traits.

There are some cases, however, when individual motion is of completely different type and can by no means be considered as random. One example is given by species migration when species transport seems to be more adequately described by Eq. (1.21) rather than by the diffusion equation. In Section 2.1 we provide a more detailed consideration of different types of individual motion and in Section 4.1 we will reveal how the interplay between the different types or modes of species transport may affect the pattern and rate of biological invasion.

In a more general situation, species transport takes place due to a combination of environmental and biological factors. In particular, individuals of a given species can be carried by the wind or current of fixed direction and strength and also be involved in random dispersal either due to self-motion or due to the impact of turbulent pulsations, cf. (Okubo, 1980). In this case, instead of Eq. (1.21) we have

$$\frac{\partial u}{\partial t} + A_0 \frac{\partial u}{\partial x} = \frac{\partial}{\partial x}\left(D\frac{\partial u}{\partial x}\right) \tag{1.22}$$

where D quantifies either biodiffusion or turbulent diffusion, respectively.

Another mechanism of transport is originated in the impact of "vector" species, e.g., in the case when plant seeds are borne away by birds or animals. Its main features are determined by the properties of individual self-motion of the vector species and, thus, the above arguments apply to this case as well. This mechanism has been proved to enhance dispersal of some plant species, although its importance is often regarded as insignificant compared to wind-driven transport (but see Section 8.3).

Finally, species transport can be caused by some anthropogenic factors such as trade, tourism, etc. From the modeling standpoint, this mechanism seems to be the most complicated because it is obviously affected by various aspects of human life and society.

Although all the mechanisms described above potentially lead to the species spatial re-distribution, sometimes on a global scale, not all of them can lead to species invasion or colonization. The factors that make a given mechanism relevant to biological invasion as well as possible implication of different types of species transport will be discussed in the next section.

1.4 Biological invasion: main facts and constituting examples

The term "biological invasion" is a common name for a variety of phenomena related to introduction and spread of alien or exotic species, i.e., a species that has not been present in a given ecosystem before it is brought in. Biological invasion usually has dramatic consequences for the native ecological community. Invasion of alien species often results in virtual eradication of some native species, and it is currently considered as one of the main reasons for biodiversity loss all over the world. It often causes considerable damage to agriculture (or to "aquaculture," in case of marine ecosystems) and thus it may result in substantial economic losses as well.

Biological invasion is a very frequent phenomenon and its frequency has increased significantly over the last several decades, although not all cases are documented equally well. A classical example is found in the spread of muskrat (*Ondatra zibethica*) in Central Europe in the first half of the 20th century; this case was later used by Skellam (1951) for developing one of the first-ever mathematical approaches to modeling species spatial spread. The muskrat was brought to Europe from North America for the purpose of fur-breeding. In 1905, a few muskrats escaped from a farm near Prague. This small population started multiplying and increasing its range, and in a few decades it spread over the whole continental Europe.

Another famous example of biological invasion is the spread of the gypsy moth (*Lymantria dispar*) in North America. This insect is thought to have been brought from France to a place near Boston by an amateur entomologist. When it happened to escape around 1870, it established a local population and eventually started spreading. In spite of a number of controlling measures introduced in the early 20th century, it has continued spreading ever since and by 1990 the whole Northwest of the United States became heavily infested (Liebhold et al., 1992). The resulting damage to agriculture and corresponding economic losses were tremendous.

One of the latest cases is given by the introduction and spread of the zooplankton species *Mnemiopsis leidyi* in the Black Sea (Vinogradov et al., 1989). This species was accidentally brought by cargo ships from the Caribbean Sea region and released with ballast waters near the port of Odessa around 1980. The species successfully established a local population but its spatial spread did not begin until mid-eighties. From spring 1988, however, the population started growing and spreading at a very high rate and it invaded the whole sea in just a few months. The consequences for the native fauna were catastrophic; in particular, a few commercial fish species were brought to the edge of extinction.

Analysis of these and many other cases shows that biological invasion has a few more or less clearly distinguishable stages. The first one is introduction when a number of individuals of an exotic species is brought, accidentally or deliberately, into a given ecosystem. The second stage is establishment when the introduced species is "getting accustomed" to the new environment. (Note that, at this stage, it is not enough just to have a number of adult individuals who survived their introduction. An alien species can be regarded as established only after offsprings are procreated in the new environment and the population starts growing or, at least, stabilizes at a certain level.) The third stage, in case the previous two have been successful, is related to the species geographical spread when the species range grows steadily and it invades new areas at the scale much larger than the place of its original introduction.

Apparently, geographical spread of alien species is only possible if there exists an adequate mechanism of species spatial re-distribution. Not every mechanism of species transport can result in biological invasion. Remarkably, the magnitude of the travel distance characteristic for a given type of transport has very little to do with its relevance to species invasion. For instance, recurrent long-distance migrations typical for many bird species usually take place without changing species range: a flock of birds may travel a thousand miles in order to spend winter in a warmer place and every next spring it returns to exactly the same forest or pond to produce new offspring. On the contrary, the everyday roundabout motion of individuals that takes place on a much smaller scale can result in a gradual increase in the species range and thus make species invasion possible.

Also, the implication for species invasion is different for different types of transport. While random self-motion of individuals and small-scale migrations (cf. Chapter 4) lead to a gradual advance of the species' range border, the impact of tourism and trade is likely to result in new introductions far away from the area already invaded.

Each stage of biological invasion has its own specific problems and requires specific modeling approaches. In this book, we are mostly concerned with the third and, to some extent, the second stage when the size of invasive population becomes large enough to be adequately described in terms of the population density and the spread can be described by diffusion-reaction equations.

It should be mentioned here that an increase in species range and corresponding geographical spread occurs not only during biological invasion but also during species colonization or recolonization. The biological background of these phenomena is somewhat different and, in the latter cases, the introduction and establishment stages may be not applicable. The geographical spread, however, takes place due to essentially the same mechanisms and can be studied using the same modeling techniques, e.g., see Lubina and Levin (1988).

In general, biological invasion is a complex phenomenon and it has many different aspects and implications. It is not our goal to give its comprehensive description here; an interested reader can find more details in relevant biological literature (Elton, 1958; Drake et al., 1989; for a more recent source see Sakai et al., 2001). The mathematical background and a variety of modeling approaches is shown and discussed in much detail in the already-classical books by Hengeveld (1989) and Shigesada and Kawasaki (1997). Some more references will be given throughout this book.

Chapter 2

Models of biological invasion

In this chapter, we will briefly review the main mathematical approaches that are used to describe biological invasions and species spread. Spatiotemporal models of population dynamics arise when the trade-off is considered between the change in population size or density due to the processes of birth and death and its change due to emigration/immigration into a given site or into a vicinity of certain position in space. There are, however, different ways to describe this trade-off depending on peculiarities of ecological situation as well as different features of population dynamics to be taken into account. This results in a wide variety of relevant mathematical models ranging from fully deterministic to purely stochastic, from single-species to multi-species, from continuous to discrete, etc. Some of them are presented below.

2.1 Diffusion-reaction equations

Application of diffusion-reaction models to ecological problems is based on the assumption that population dynamics is continuous in space and time and thus can be described by continuous or even smooth functions. Apparently, this approach fully neglects population discreteness at the level of individuals and thus can only be applied to the processes going on a spatial scale much larger than the size of a typical individual.

Let us consider how the population size changes inside a certain area Ω. We assume that a given population is described by the population density u. In a general case, u depends on the position $\mathbf{r} = (x, y)$ in space and time t so that the population size inside Ω is given by

$$U_\Omega(t) = \int\int_\Omega u(\mathbf{r}, t) d\mathbf{r} \ . \tag{2.1}$$

There are two basically different mechanisms making the population size vary with time: the one associated with local processes such as birth, death, predation etc., and the other associated with the redistribution of the population in space due to the motion of its individuals. Correspondingly, the rate of change of the population size is described by the following equation:

$$\frac{\partial}{\partial t}\int\int_{\Omega} u(\mathbf{r},t)d\mathbf{r} = -\int_{\Gamma}(\mathbf{Jk})ds + \int\int_{\Omega} F(u(\mathbf{r},t))d\mathbf{r} \tag{2.2}$$

where $\mathbf{J}$ is the population density flux through the area boundary Γ, $\mathbf{k}$ is the outward-pointed unit vector normal to the boundary and $\mathbf{Jk}$ is the scalar product. The second term in the right-hand side allows for the local processes, $F(u) = f(u)u$ where $f(u)$ is the per capita growth rate.

Taking into account that

$$\int_{\Gamma}(\mathbf{Jk})ds = \int\int_{\Omega}(\nabla\mathbf{J})\, d\mathbf{r}\ , \tag{2.3}$$

from (2.2), we obtain:

$$\frac{\partial u(\mathbf{r},t)}{\partial t} = -\nabla\mathbf{J} + F(u). \tag{2.4}$$

In the case $F(u) \equiv 0$, Eq. (2.4) has the form of the conservation law; it means that the total number of the individuals does not change with time unless the processes of birth and death are taken into account.

The form of the flux $\mathbf{J}$ essentially depends on the properties of the motion. In case the motion of the individuals can be regarded as random, cf. "random walk" (Okubo, 1980), the flux is usually assumed to be proportional to the population density gradient:

$$\mathbf{J} = -D\nabla u(\mathbf{r},t) \tag{2.5}$$

where D is diffusivity. In other fields of natural science, equation (2.5) is known as the Fick law or the Fourier law. In this case, Eq. (2.4) takes the form of a diffusion-reaction equation which is of common use in theoretical studies (Britton, 1986; Murray, 1989; Holmes et al., 1994).

However, the motion of individuals cannot always be regarded as random. Another widely observed dynamics is advection/migration, when the individuals exhibit a correlated motion toward a certain direction. Assuming for the sake of simplicity that at a given position all the individuals move with the same speed $\mathbf{A}$, we immediately obtain that $\mathbf{J} = \mathbf{A}u(\mathbf{r},t)$. In a more general case, when the correlated motion is combined with the random motion, the population density flux is given by the following equation:

$$\mathbf{J} = \mathbf{A}u(\mathbf{r},t) - D\nabla u(\mathbf{r},t)\ . \tag{2.6}$$

From (2.4) and (2.6), we obtain the following general advection-diffusion-reaction equation of population dynamics:

$$\frac{\partial u(\mathbf{r},t)}{\partial t} + \nabla(\mathbf{A}u) = \nabla\left(D\nabla u\right) + f(u)u\ . \tag{2.7}$$

Here the diffusion coefficients D and the advection speed $\mathbf{A}$ may depend on space, time and also on population density u.

The above derivation of Eq. (2.7) can be immediately extended onto the case of a few interacting species. In this case, instead of a single equation, we arrive at the system of advection-diffusion-reaction equations:

$$\frac{\partial u_i(\mathbf{r},t)}{\partial t} + \nabla(\mathbf{A}_i u_i) = \nabla\left(D_i \nabla u_i\right) + f_i(u_1,\ldots,u_n)u_i, \qquad i = 1,\ldots,n \tag{2.8}$$

where u_i is the density of the ith species and the nonlinear functions f_i describe the inter-species interactions. Here diffusivity is, in general, different for different species. Also, the speed of advective transport can be species-specific in case we take into account self-motion of individuals.

System (2.8) is very general and, as such, it can be expected to describe a great variety of different situations in various natural systems. However, from a practical point of view, a disadvantage of this model is that it is mathematically very complicated. Its numerical solution is usually a difficult problem, and analytical approaches are hardly possible at all except for a few special cases.

Moreover, the idea to include as many details as possible, resulting in a large number n of equations in the system, is not always justified biologically because complicated systems tend to be sensitive to parameter values that are usually known only approximately. On the contrary, in many cases it appears possible to make useful insights into the population dynamics of invasive species without describing inter-species interactions in much detail, i.e., using few-species models. A few ecologically meaningful examples will be considered in Chapter 8. An apparent advantage of the few-species models is that they can be more easily solved numerically and often treated analytically.

The simplest case is a single-species model, i.e., the one given by equation (2.7). Although this model does not take into account the impact of other species (but see the lines below Eq. (2.11)), it still takes into account several different factors such as two types of motion, potential density-dependence, environmental heterogeneity (in case D, $\mathbf{A}$ and/or other parameters depend on space), etc. Depending on the nature of phenomenon under study, the model (2.7) can be further minimized. In most of this book we neglect advection, its impact will be considered in Section 4.1. Also, it is often possible to treat the diffusion coefficient as constant. Under these additional assumptions, we arrive at the following equation:

$$\frac{\partial u(\mathbf{r},t)}{\partial t} = D\nabla^2 u + F(u) \ . \tag{2.9}$$

Analogously, a similar reduction can be applied to the general advection-diffusion-reaction system (2.8) as well.

Some conclusions about the solution properties for the above equation can be obtained for functions $F(u)$ of a rather general form; however, an exact solution is possible only when a specific parameterization is chosen. The choice of $F(u)$ is governed by biological reasons. For instance, in case the population growth is subject to the Allee effect, a cubic polynomial can be an appropriate parameterization; see (1.16). Equation (2.9) then takes a more specific form:

$$\frac{\partial u(\mathbf{r},t)}{\partial t} = D\nabla^2 u + \gamma u(u-\beta)(1-u) \tag{2.10}$$

assuming that population density u is scaled by the population carrying capacity, cf. Section 1.2.

In the case that the Allee effect is absent, the population growth is often described by the logistic function; correspondingly, equation (2.9) turns into

$$\frac{\partial u(\mathbf{r},t)}{\partial t} = D\nabla^2 u + \alpha u(1-u) \ . \tag{2.11}$$

Equation (2.11)) was first introduced by Fisher (1937); for that reason it is called the Fisher equation.

Note that, although at first sight it may seem that Eq. (2.9) (or its particular cases (2.10) and (2.11)) fully neglects the interaction of the given population with other species, it is not really so. The term $F(u)$ may include the impact of other species in an implicit way, e.g., via additional mortality. Still, this approach provides only meagre information about other species' impact on given population and gives no information at all about how other species can be affected. A more comprehensive model should take other species into account explicitly. In particular, a lot of important results and valuable insights have been obtained from consideration of the two-species predator-prey system:

$$\frac{\partial u(\mathbf{r},t)}{\partial t} = D_1\nabla^2 u + f(u)u - r(u)uv \ , \tag{2.12}$$

$$\frac{\partial v(\mathbf{r},t)}{\partial t} = D_2\nabla^2 v + \kappa r(u)uv - g(v)v \tag{2.13}$$

where u, v are the densities of prey and predator, respectively, the term $r(u)v$ stands for predation, κ is the coefficient of food utilization and $g(v)$ is the per capita mortality rate of predator.

Throughout this book, most of our analysis will be restricted to the 1-D case (but see Sections 4.2 and 8.1) when the corresponding equations read as follows:

$$u_t(x,t) = Du_{xx} + F(u) \tag{2.14}$$

for a single-species model, and

$$u_t(x,t) = D_1u_{xx} + f(u)u - r(u)uv \ , \tag{2.15}$$

$$v_t(x,t) = D_2v_{xx} + \kappa r(u)uv - g(v)v \tag{2.16}$$

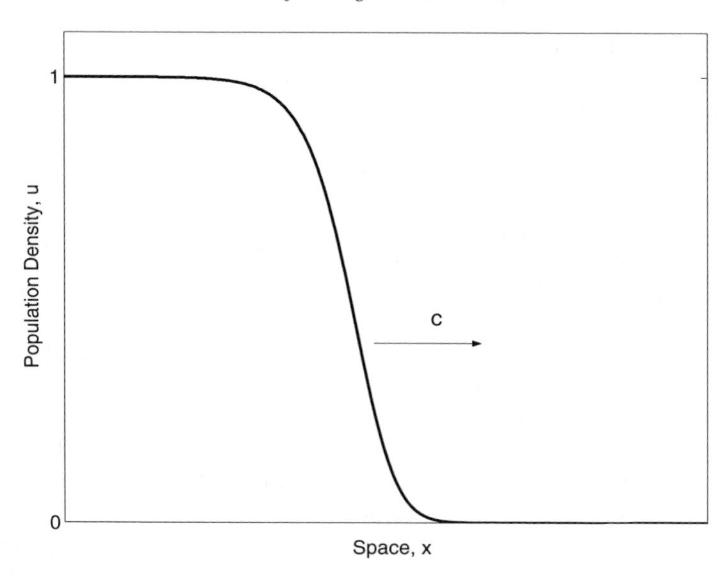

FIGURE 2.1: A propagating population front described by a diffusion-reaction equation.

for a predator-prey model where indices x and t denote the partial derivatives with respect to x and t, respectively.

Regarding biological invasions, the spread of alien species usually takes place via propagation of a population front separating the regions where the invasive species is absent from the regions where it is present at a considerable density. Therefore, the solutions that describe propagation of traveling waves are of particular interest. The relevant solutions depend not on x and t separately but on the special combination of space and time, $u(x,t) = U(\xi)$ where $\xi = x - ct$ and c is the speed of the wave. In the single-species system, obviously, they are solutions of the following equation:

$$D\frac{d^2U}{d\xi^2} + c\frac{dU}{d\xi} + F(U) = 0 \ . \tag{2.17}$$

In an unbounded domain, propagation of a traveling front corresponds to the following conditions at infinity:

$$U(-\infty) = 1, \qquad U(\infty) = 0 \tag{2.18}$$

(or vice versa, depending on the direction of axis x).

For the growth function given by conditions either (1.8–1.10) or (1.13–1.14), the only possible type of traveling wave described by Eq. (2.17) appears to be a propagating front connecting the lower and the upper steady states, i.e., $u = 0$ and $u = 1$; see Fig. 2.1. The speed of the wave strongly depends on the type of function F which, in its turn, is determined by biological factors.

Assuming that the local population growth does not exhibit the Allee effect, i.e., that $F(u)$ is described by conditions (1.8–1.10), the following analytical expression for the wave speed can be obtained:

$$c = 2\,(D\alpha)^{1/2} \tag{2.19}$$

where $\alpha = F'(0)$, cf. (2.11). More rigorously, Eq. (2.19) gives the lowest possible value of the wave speed (see Chapter 7 for more details); however, it is this value that appears to be relevant to the population waves arising in biological invasion.

The situation becomes essentially different when the population growth is subject to the Allee effect. In this case, equation (2.19) does not apply. Considerations of this issue (Volpert et al., 1994) show that the speed of the wave depends on more details of the function $F(u)$ than just its behavior at small u. In fact, an exhaustive solution of this problem for a more or less general case is still lacking; however, for a few special cases the equation for the wave speed has been obtained. In particular, when the local population growth is described by a cubic polynomial, cf. Eqs. (1.16) and (2.10), the speed is given as follows:

$$c = c_0\,(D\gamma)^{1/2} \quad \text{where} \quad c_0 = \frac{1-2\beta}{\sqrt{2}}\ . \tag{2.20}$$

Remarkably, depending on β, the direction of population front propagation can be different and corresponds either to species invasion or to species retreat.

The question of primary importance for ecological applications is under what conditions the wave of invading species can be blocked. For the single-species model (2.14) or (2.17), this condition can be easily obtained. Wave blocking means $c = 0$; correspondingly, from Eq. (2.17) we have:

$$D\frac{d^2U}{d\xi^2} + F(U) = 0\ . \tag{2.21}$$

Multiplying (2.21) by $dU/d\xi$ and integrating over space, we obtain:

$$\int_0^1 F(U)dU = 0 \tag{2.22}$$

(assuming, as above, that U is scaled to unity). A certain generalization of condition (2.22) will be given in Section 5.2 (see also Chapter 7).

Note that condition (2.22) can hold only if the function F changes its sign in the interval $(0,1)$. Apparently, this is not possible in the case when the population growth is logistic (see (1.8–1.10)), but it appears possible in the case when the population growth is damped by the strong Allee effect, i.e., when $F(u)$ is negative for small u. In the latter case, the condition of wave blocking has the clear geometrical interpretation that the areas above and below the horizontal axis must be equal; see Fig. 2.2.

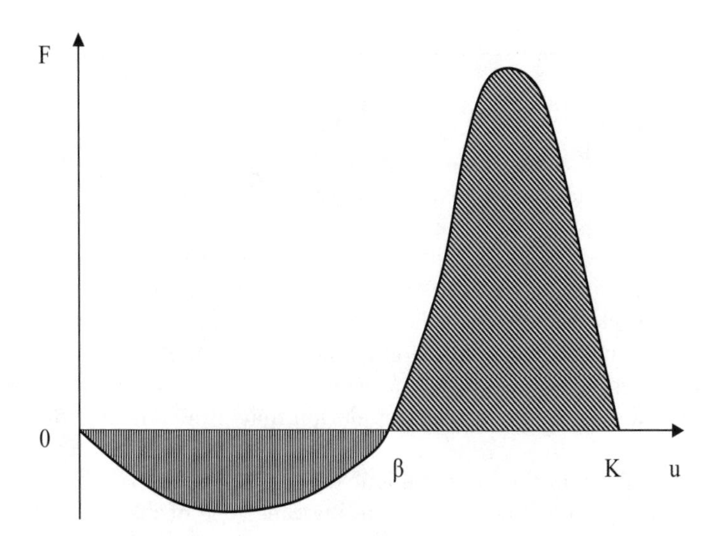

FIGURE 2.2: Geometrical interpretation of the condition of wave blocking in the population with the strong Allee effect: species invasion is stopped when the shadowed areas above and below the horizontal axis are equal.

In the case of the strong Allee effect, in the parameter range where $\int F(u)du$ is close to zero, the speed of the wave was shown to be given as

$$c \simeq \int_0^1 F(u)du \;, \tag{2.23}$$

see Mikhailov (1990). This and similar results make it possible to refer to the traveling fronts in the systems with and without Allee effect as "pushed" and "pulled" waves, respectively: while in the case of logistic growth the speed of the population wave depends on the system properties at the leading edge, i.e., far in front of the front where the population density is small, in the case of the Allee effect the speed of the wave depends on the properties of F for intermediate and large population density, i.e., behind the front.

It should be mentioned that the above mechanism of wave blocking due to the impact of the Allee effect appears to be the only one possible in the "minimal" model (2.14). In more complicated diffusion-reaction models, wave blocking appears possible as well due to the impact of other factors such as environmental heterogeneity and impact of other species.

2.2 Integral-difference models

Diffusion-reaction equations of population dynamics, although proved to be very useful in many ecological applications, e.g., see Chapter 8, apparently give an idealized picture of real population dynamics. In particular, a point of criticism has been that, in diffusion-reaction models, the individuals move in space and multiplicate at the same time. In reality, for many species these stages can be clearly distinguished. For instance, re-distribution in space (including also invasion/colonization phenomena) of a plant population takes place due to seed or pollen dispersal which normally happens once per year. Between the dispersal events, the population may grow but its individuals do not move.

In an attempt to find a more realistic description, in particular, in order to take into account the stage-separation, another type of mathematical model was developed based on integral-difference equations. In this section we will make a brief review of this approach and try to compare its predictive ability with that of diffusion-reaction models. For the sake of equations' simplicity, we restrict our consideration to 1-D case and to population with non-overlapping generations. We also assume that dispersal takes place in an infinite space in order to neglect the impact of the area boundaries. More details and some generalizations as well as further references can be obtained from Kot et al. (1996), Neubert and Caswell (2000) and Wang et al. (2002).

Let us describe a given population with its density u which depends on time t and position x. Function $u(x,t)$ gives the population distribution over space at time t. Then, the population density at time $t+\Delta t$, i.e., $u(x,t+\Delta t)$, is the result of the following two processes: the population growth described by the growth function $\mathcal{F}(u)$ and redistribution in space due to offsprings' dispersal.

The actual mathematical model depends on the succession of these processes. Let T be the time between the two subsequent generations. Assuming that the species first multiplicates then disperse, we arrive at the following two-step model, i.e.,

$$\tilde{u}(x,t) = \mathcal{F}(u(x,t)) \tag{2.24}$$

for the multiplication stage, and

$$u(x,t+T) = \int_{-\infty}^{\infty} k(x,y)\tilde{u}(y,t)dy \tag{2.25}$$

for the dispersal stage. Here $k(x,y)$ is the dispersal kernel giving the probability density to find at position x an offspring released at position y.

Equations (2.24-2.25) can be combined together producing a single equation:

$$u(x,t+T) = \int_{-\infty}^{\infty} k(x,y)\mathcal{F}(u(y,t))dy\ . \tag{2.26}$$

This equation, or its immediate generalizations, has been often used in applied ecological studies, cf. Clark et al. (1998).

Now, let us note that we can assume as well that the species first disperse and then reproduce. Then, we immediately arrive at the alternative two-step model:

$$\tilde{u}(x,t) = \int_{-\infty}^{\infty} k(x,y)u(y,t)dy \tag{2.27}$$

for the dispersal stage, and

$$u(x,t+T) = \mathcal{F}(\tilde{u}(x,t)) \tag{2.28}$$

for the multiplication stage. Again, we can combine the equations (2.27-2.28); however, the resulting equation will be different:

$$u(x,t+T) = \mathcal{F}\left(\int_{-\infty}^{\infty} k(x,y)\tilde{u}(y,t)dy\right). \tag{2.29}$$

From a biological point of view, equations (2.26) and (2.29) should be equivalent because the actual succession of stages is only a subject of the choice of the initial conditions. Assuming that, as it is usually the case in nature, the system exhibits some kind of scaling so that the initial conditions are "forgotten" after a certain time, we arrive at the conclusion that $u(x,t+\Delta t)$ in equations (2.26) and (2.29) should be the same, i.e.,

$$\int_{-\infty}^{\infty} k(x,y)\mathcal{F}(u(y,t))dy = \mathcal{F}\left(\int_{-\infty}^{\infty} k(x,y)u(y,t)dy\right). \tag{2.30}$$

Equation (2.30) expresses a "stage-invariance" principle: the large-time dynamics of given population must not depend on which stage, dispersal or multiplication has taken place first.

It is readily seen that it is only possible when $\mathcal{F}(u)$ is a linear function, i.e., $\mathcal{F}(u) = R_0 u$ where R_0 is the basic reproductive number giving an average number of offsprings produced by a single adult individual. Thus, in a rigorous mathematical sense, the approach based on Eq. (2.26) is true only when the density-dependent phenomena in population dynamics are neglected. Although it is a serious restriction for its applications because density-dependence often plays a crucial role, surprisingly, equation (2.26) with $\mathcal{F}(u) = R_0 u$ gives a good estimate for the invasion speed. The matter is that, in the case that an invasive species spreads in the form of a traveling population wave, in the absence of the Allee effect the wave speed is determined by the population dynamics at the leading edge of the moving front (Weinberger, 1982). At the edge of the front, the population density is small and thus function $\mathcal{F}(u)$ can be linearized, $\mathcal{F}(u) \simeq \mathcal{F}'(0)u$.

For further analysis, we assume that the environment is homogeneous so that dispersal probability depends on the distance between positions x and y

rather than on both of them separately, i.e., $k(x,y) = k(x-y)$. Then, in the linear case, equation (2.26) takes a simpler form:

$$u(x,t+T) = R_0 \int_{-\infty}^{\infty} k(x-y)u(y,t)dy \ . \tag{2.31}$$

A stationary wave (i.e., the traveling wave with a profile of a constant shape) traveling with a speed c is a solution which exhibits invariance to translation, i.e., in the case of (2.31), possesses the following property:

$$u(x,t+T) = u(x-cT,t) \ . \tag{2.32}$$

Since equation (2.31) is linear, we can look for its solution to be exponential, i.e.,

$$u(x,t) = \text{const} \cdot e^{-s(x-ct)} \tag{2.33}$$

where $s > 0$ (assuming that the invasive species spreads from left to right) is a parameter giving the slope of the front.

Having substituted (2.33) into (2.31), we obtain:

$$e^{-sx}e^{scT} = R_0 \int_{-\infty}^{\infty} k(x-y)e^{-sy}dy \ . \tag{2.34}$$

Introducing $z = x - y$, we arrive at the characteristic equation:

$$e^{scT} = R_0 M(s) \ \text{ where } \ M(s) = \int_{-\infty}^{\infty} k(z)e^{sz}dz \ . \tag{2.35}$$

From (2.35), we immediately obtain the equation for the speed of the population wave:

$$c(s) = \frac{1}{sT}\ln[R_0 M(s)] \ . \tag{2.36}$$

It is readily seen that equation (2.36) has multiple solutions corresponding to different s. Which value of speed is actually "chosen" by the traveling wave depends on the initial spatial distribution of species density, in particular, on its asymptotics for large x. However, biological invasion usually starts with species local introduction so that before starting its geographical spread the exotic species is present only inside a certain area. Correspondingly, relevant initial conditions are described by functions of compact support. In this case, it can be proved (Weinberger, 1982) that the wave propagates with the minimum value given by equation (2.36). Thus, the relevant value of the invasion speed is given by the following equation:

$$c_0 = \min_s \frac{1}{sT}\ln[R_0 M(s)] \ . \tag{2.37}$$

The above analysis was based on the assumption that the moment-generating function $M(s)$ exists, at least, for certain positive values of s. That implies that, for large $|z|$, $k(z)$ decays exponentially or faster. In the case of "fat-tailed" kernels when $M(s)$ is infinite, a more refined analysis predicts that the invasive species spreads with increasing speed; see Kot et al. (1996).

Based on the integro-difference equation (2.26), Kot et al. (1996) fulfilled computer simulations of the spread of invasive species using the data of field observations on *Drosophila pseudoobscura*. The dispersal kernel was chosen to approximate the real data of the insects spread from a point release. They tested a few parameterizations and found, among other results, that leptokurtic kernels (e.g., "back-to-back" exponential) tend to yield higher speed of invasion than the standard Gaussian kernel. The model capability to give higher speed of invasion was then interpreted as an essential advantage of the integro-difference equations compared to other models.

It should be mentioned that, although from a biological point of view integro-difference equations may look somewhat more realistic compared to diffusion-reaction equations because of the separation between the dispersal and multiplication stages (but see the "stage-invariance" principle above), their better predictive ability seems to have been somewhat overestimated – at least in the case when the dispersal kernel is not fat-tailed. In particular, below we show that different speed of invasion for different dispersal kernels can also be obtained for diffusion-reaction equations.

To address this issue, let us first try to reveal the relation between integral-difference and diffusion-reaction equations. For the sake of simplicity, here we focus on the dispersal stage and neglect species multiplication:

$$u(x, t + \Delta t) = \int_{-\infty}^{\infty} k(x - y)u(y, t)dy \tag{2.38}$$

where Δt is small compared to generation time T. Let $z = x - y$, then

$$u(x, t + \Delta t) = \int_{-\infty}^{\infty} k(z)u(x - z, t))dz \tag{2.39}$$

$$= \int_{-\infty}^{\infty} k(z) \left[u(x, t) - \frac{du}{dz} \cdot z + \frac{d^2u}{dx^2} \cdot \frac{z^2}{2} - \frac{d^3u}{dx^3} \cdot \frac{z^3}{6} + \ldots \right] dz$$

assuming, for biological reasons, that $u(x, t)$ is an analytical function and thus can be expanded into a power series.

To get a step further, we also have to assume existence of all moments of the dispersive kernel $k(z)$, i.e.,

$$\int_{-\infty}^{\infty} z^n k(z)dz < \infty \ , \quad n \ = \ 1, 2, \ldots \ . \tag{2.40}$$

Obviously, conditions (2.40) are consistent with the earlier assumption about existence of the moment-generating function $M(s)$.

Then, taking into account that

$$\int_{-\infty}^{\infty} k(z)dz = 1 \tag{2.41}$$

since $k(z)$ gives the probability distribution, from (2.39) we obtain:

$$u(x,t+\Delta t) = u(x,t) - <z> \frac{du}{dz} \tag{2.42}$$

$$+ \left\langle \frac{z^2}{2} \right\rangle \frac{d^2u}{dx^2} - \left\langle \frac{z^3}{6} \right\rangle \frac{d^3u}{dx^3} + \ldots$$

where

$$\left\langle \frac{z^n}{n!} \right\rangle = \int_{-\infty}^{\infty} \frac{z^n}{n!} k(z)dz \ . \tag{2.43}$$

Assuming that there is no prevailing wind or current, the offspring dispersal forth and back must take place with equal probability; that means that $k(-z) = k(z)$ and all odd moments turn to zero. Then from (2.42) we obtain:

$$\Delta u = \left\langle \frac{z^2}{2} \right\rangle \frac{d^2u}{dx^2} + \left\langle \frac{z^4}{24} \right\rangle \frac{d^4u}{dx^4} + \ldots \tag{2.44}$$

where $\Delta u = u(x,t+\Delta t) - u(x,t)$ is the change in population density at position x after time Δt.

Since $\Delta u \approx u_t(x,t)\Delta t$, equation (2.44) has a structure similar to the diffusion equation. A formal distinction is that, although for many kernels the higher moments can be expected to be small, equation (2.44) contains higher derivatives and thus is not equivalent to the diffusion equation. Let us note, however, that, in order to obtain equation (2.44), we did not take into account that the dispersal kernel (more precisely, its width and amplitude) should depend on Δt. Clearly, the greater Δt is, the more offsprings will be found at larger distances from their parent. That means that the kernel's width and amplitude must be increasing and decreasing functions of Δt, respectively.

In order to find out how the kernel's width may depend on Δt, we can make use of dimension analysis. Since the kernel k is unknown and, moreover, it is likely to be different for different species, the function $k(x-y,\Delta t)$ should depend on a dimensionless combination of variables and parameters rather than on each of them separately. (Otherwise, what could it be, for instance, logarithm of 1 meter?) Let us introduce an empirical value of species diffusivity D so that $\sqrt{D}$ gives the average distance travelled by the offsprings per unit time from the point of their release. Since the dimension of D is L^2T^{-1}, the combination $\xi = (x-y)(D\Delta t)^{-1/2}$ is dimensionless. Then, the dispersal kernel should have the following properties:

$$k(x-y,\Delta t) = A\tilde{k}\left(\frac{x-y}{2\sqrt{D\Delta t}}\right), \qquad \int_{-\infty}^{\infty} \tilde{k}(\xi)d\xi = 1 \tag{2.45}$$

where A is a coefficient and the coefficient 2 in the denominator is added for convenience. To obtain A, we make use of (2.41):

$$A\int_{-\infty}^{\infty}\tilde{k}\left(\frac{x-y}{2\sqrt{D\Delta t}}\right)dy = A(4D\Delta t)^{1/2}\int_{-\infty}^{\infty}\tilde{k}(\xi)d\xi \qquad (2.46)$$

$$= A(4D\Delta t)^{1/2} = 1$$

so that $A = (4D\Delta t)^{-1/2}$.

Correspondingly, what kind of dependence on Δt will the coefficients $< z^n/n! >$ have? From (2.46), we immediately obtain:

$$\left\langle\frac{z^{2n}}{(2n)!}\right\rangle = \frac{1}{2\sqrt{D\Delta t}}\int_{-\infty}^{\infty}\frac{z^{2n}}{(2n)!}\tilde{k}\left(\frac{z}{2\sqrt{D\Delta t}}\right)dz \qquad (2.47)$$

$$= \left(2\sqrt{D\Delta t}\right)^{2n}\int_{-\infty}^{\infty}\frac{\xi^{2n}}{(2n)!}\tilde{k}(\xi)d\xi$$

$$= (4D\Delta t)^n\left\langle\frac{\xi^{2n}}{(2n)!}\right\rangle_*, \quad n = 1,2,\ldots$$

where the brackets with asterisk denote an average with respect to the probability distribution function $\tilde{k}$ of the re-scaled argument ξ.

Now we return to equation (2.44). Since, during the dispersal stage, the process of offsprings spread is more likely to happen in a continuous manner than in a discrete one, we assume Δt to be small. Form (2.44), we obtain:

$$\Delta u = u(x,t+\Delta t) - u(x,t) = u_t(x,t)\Delta t + o(\Delta t)$$

$$= 4D\Delta t\left\langle\frac{\xi^2}{2}\right\rangle_* u_{xx} + (4D\Delta t)^2\left\langle\frac{\xi^4}{24}\right\rangle_* u_{xxxx} + \ldots . \qquad (2.48)$$

Dividing Eq. (2.48) by Δt and considering $\Delta t \to 0$, we obtain:

$$u_t(x,t) = \left[\left\langle 2\xi^2\right\rangle_* D\right]u_{xx} \qquad (2.49)$$

where all higher derivatives have now disappeared.

An important distinction of equation (2.49) from a usual diffusion equation is that the diffusion coefficient in (2.48) contains the factor determined by the properties of the kernel, i.e., $< 2\xi^2 >_*$. Since the speed of invasion in a single-species diffusion-reaction model is proportional to the square root of the diffusion coefficient (e.g., see Eq. (2.19)), it may be expected that it will be different for different kernels.

Indeed, it is straightforward to see that in the standard case of Gaussian kernel, i.e., for

$$\tilde{k}(\xi) = \frac{1}{\sqrt{\pi}}\exp(-\xi^2) \qquad (2.50)$$

this factor is equal to 1. However, in the case of leptokurtic kernel,

$$\tilde{k}(\xi) = \frac{1}{2}\exp(-|\xi|) \tag{2.51}$$

this factor is equal to 4. Comparison between these two cases leads to an immediate conclusion that the speed of invasion must be two times greater in the leptokurtic case than in the Gaussian case. This result is in excellent agreement with Kot et al. (1996) where it was obtained based on the integro-difference equation (2.26).

Thus, we can conclude that, in the case that all the moments exist (even if the kernel is not Gaussian), the prediction of the invasion speed obtained from the integro-difference equation is essentially the same as that obtained from the diffusion-reaction equation. However, the models based on integro-difference equations do provide a valuable extension to diffusion-reaction equations in the case of fat-tailed kernels, i.e., when some or all of the moments are infinite, predicting population waves propagating with increasing speed.

2.3 Space-discrete models

The models considered in the previous sections are based on the assumption that the factors affecting the population dynamics, such as the population growth rate(s), mortality rate(s), diffusivity, etc., are homogeneous in space. In its turn, this assumption is based on the hypothesis that the relevant environmental properties are homogeneous as well. Apparently, this is not always true and, although spatially homogeneous models often provide an adequate description of invasive populations, in some cases environmental heterogeneity cannot be neglected.

There are various ways to take environmental heterogeneity into account. Probably the most straightforward one is simply to change constant parameters to relevant functions of position in space in the space-continuous equations. For instance, considering the carrying capacity K and/or the population growth rates α and γ in Eqs. (2.10) and (2.11) as functions of space, one takes into account the well-known ecological observation that population habitats can be either favorable (large K, large α or γ) or unfavorable (small K, small α or γ).

In nature a species often dwells in fragmented habitats with a "mosaic" structure when favorable areas alternate with areas where given species cannot survive (but probably can disperse through). For such cases, it is likely to be more convenient to consider the population dynamics on a spatially-discrete grid rather than in a continuous space with environmental properties described by complicated functions.

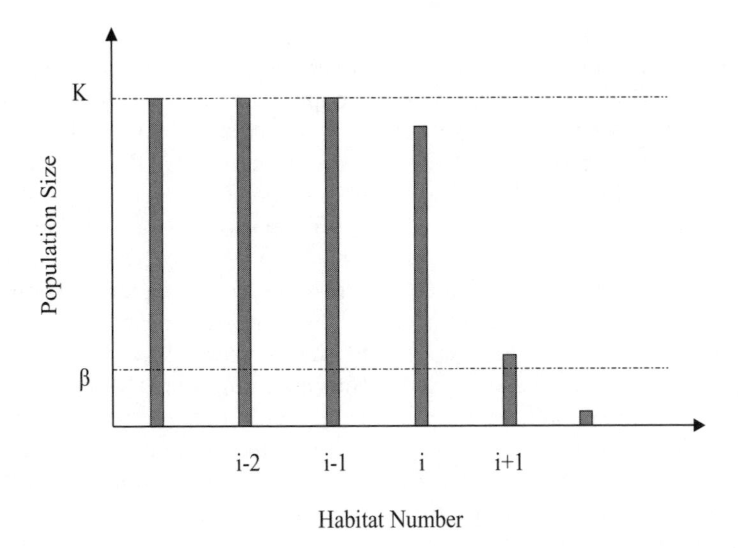

FIGURE 2.3: A sketch of an invasive population spreading through a one-dimensional system of coupled discrete habitats.

For the sake of brevity, we restrict our consideration to the single-species dynamics. We begin with a one-dimensional case. Let us consider a chain of favorable sites, or habitats, numbered consecutively; see Fig. 2.3. The sites are connected via dispersion of the individuals. The population size U_i in the i-th habitat can change due to processes of local population growth and mortality, and due to migration to/from other habitats. We assume that dispersal has only short-distance mode so that migration takes place only between the neighboring sites. We also make a usual assumption that the inter-site population flux is proportional to the difference between the corresponding population sizes. Thus, the equations describing population functioning in a fragmented environment are as follows:

$$\frac{dU_i}{dt} = J_{i,L} + J_{i,R} + F_i(U_i) \, , \quad i = 1, \ldots, N \tag{2.52}$$

where

$$J_{i,L} = D_{i,L}\left(U_{i-1} - U_i\right), \quad J_{i,R} = D_{i,R}\left(U_{i+1} - U_i\right) \tag{2.53}$$

where $D_{i,L}$, $D_{i,R}$ are the coefficients of diffusive coupling and N is the total number of sites. Here functions F_i describing local growth and mortality can be either identical or different depending on the habitats' properties. In the particular but ecologically important case of identical patches, Eqs. (2.52) reduce to

$$\frac{dU_i}{dt} = D\left(U_{i+1} + U_{i-1} - 2U_i\right) + F(U_i) \, , \quad i = 1, \ldots, N, \tag{2.54}$$

where $D_{i,L} = D_{i,R} = D$.

Note that, even in the case that all functions F_i are identical, equations (2.54) still describe population dynamics in an inhomogeneous "patchy" environment. Although Eqs. (2.54) look somewhat similar to what is obtained when a single-species diffusion-reaction equation is solved numerically by finite-difference method, these approaches are not equivalent. A system of finite-difference equations is expected to approximate a given space-continuous diffusion-reaction equation in the limiting case when the grid-steps become very small. (It means that, ideally, a diffusion-reaction equation should be solved numerically on a succession of grids with decreasing grid-steps in order to exclude numerical artifacts.) On the contrary, Eqs. (2.52–2.53) or (2.54) does not contain space explicitly thus assuming that the site sizes and inter-site distances remain finite. As a result, although the model based on the space-discrete equations (2.54) is in a certain relation to the diffusion-reaction model in a homogeneous environment, it predicts some new features of invasive species dynamics. In particular, while in the case of strong inter-site coupling the system (2.54) has traveling wave solutions describing the invasion of alien species, in the case of weak coupling (but with finite D) the system (2.54) predicts blocking of invasive species spread. The latter happens in a parameter range where it cannot be immediately described by means of the corresponding space-continuous model. Wave blocking in a space-discrete system due to weak inter-site coupling was studied mathematically by Keener (1987), Bressloff and Rowlands (1997), Fath (1998) and some others. Ecological implications of this phenomenon were considered by Keitt et al. (2001).

The existence of traveling population fronts in the space-discrete system can be demonstrated straightforwardly, at least for a certain specific choice of $F(U)$. Bressloff and Rowlands (1997) considered the following parameterization

$$F(U) = \epsilon \left[\left(1 - \frac{a}{2}\right) - U^2 \right] - \frac{aU}{1 - U^2} + 2U \tag{2.55}$$

where ϵ and a are certain parameters without any immediate biological meaning, $\epsilon > 0$, $0 \le a \le 2$. It is not difficult to see that function (2.55) possesses the following properties:

$$F(0) = F(U_{min}) = F(U_{max}) = 0 ,$$
$$F(U) < 0 \text{ for } U_{min} < U < 0 , \quad F(U) > 0 \text{ for } 0 < U < U_{max}$$

(where U_{min} and U_{max} depend on a and ϵ) so that it describes the dynamics similar to the one generated by the strong Allee effect, cf. (1.13–1.14).

Bressloff and Rowlands (1997) studied the system (2.54–2.55) and found exact traveling wave solution:

$$U_i(t) = \tanh b \cdot \tanh \left[b\,(ct - n) + s \right] \tag{2.56}$$

(in appropriately chosen dimensionless variables) where

$$\tanh^2 b = 1 - \frac{a}{2} \,, \tag{2.57}$$

s is an arbitrary constant and the speed of the wave c was found to be

$$c = \frac{\epsilon}{b} \, \tanh b \,. \tag{2.58}$$

Unfortunately, the choice of F in the form (2.55) leads to the values of U to be scaled between $U_{max} > 0$ and $U_{min} < 0$ so that U can be negative, and thus the ecological meaning of solution (2.56) remains obscure. Also, to obtain the exact solution (2.56), Bressloff and Rowlands (1997) had to assume that $D = 1$ which clearly reduce its usefulness. In spite of these apparent disadvantages of their approach, the importance of this result is that it gives a mathematically rigorous proof of traveling wave existence in a discrete environment, at least for a special case.

Thus, propagation of traveling population waves is the phenomenon that can be observed both in space-continuous models and in space-discrete models. The property that makes a principal distinction between these approaches is the parameter range where wave propagation is possible. In a single-species diffusion-reaction system, traveling wave of invasion can be blocked or reversed only in the case when the population is affected by the Allee effect. In this case, the condition of propagation failure is given by the following inequality:

$$M \; = \; \int_0^K F(U)dU \; \leq \; 0 \tag{2.59}$$

where K is the population carrying capacity.

For a specific parameterization of the growth rate $F(U)$, inequality (2.59) gives restriction on the parameter values; in particular, in case of cubic polynomial parameterization (see (2.10)), condition(2.59) is equivalent to $\beta \geq 0.5$. Moreover, since $M < 0$ actually corresponds to a reverse traveling wave, i.e., to the species retreat, the wave blocking occurs for the only value $M = 0$. Dependence of the wave speed on species diffusivity is given by the simple scaling relation $c \sim \sqrt{D}$ so that for $M \neq 0$ species invasion or retreat takes place for any $D > 0$.

In contrast, in a space-discrete system the traveling wave of invasive species can be blocked for $M > 0$ provided that inter-site diffusive coupling is sufficiently small. The possibility of this effect can be seen from simple heuristic arguments (Keitt et al., 2001). Let us consider a certain invasive species that spreads over a system of identical coupled habitats, cf. Fig. 2.3. Without any loss of generality, we assume that it spreads from left to right and it has already invaded the first $(i-1)$ sites. Successful invasion over the i-th habitat means that its population size will grow, i.e., $dU_i/dt > 0$. That becomes possible only when the negative impact of the Allee effect is overcome by a sufficiently strong population flux from the neighboring $(i-1)$-th site. The flux in question is given by the following equation:

$$J_{i,L} = D(U_{i-1} - U_i) \ . \tag{2.60}$$

The values of U_i and U_{i-1} are unknown and thus the exact value of the flux is unknown as well. However, its upper bound is obtained readily if we take into account that the value of population size in the "already-invaded" habitats cannot exceed the carrying capacity K and the population size in "not-yet-invaded" habitats is nonnegative. Then, from (2.60) we immediately obtain:

$$J_{i,L} \ < \ J_{max} \ = \ DK \ . \tag{2.61}$$

Evidently, a sufficient condition of invasive species blocking is given by

$$J_{max} + \min F(U) < 0 \tag{2.62}$$

where the minimum value is taken over the range where the population growth rate is negative, i.e., for $0 < U/K < \beta$. Under this condition, the population "source" due to the flux from the $(i-1)$-th site (more rigorously, its theoretical maximum value) is weaker than the "sink" due to the Allee effect. It means that dU_i/dt appears to be negative in a certain range of the population density, U_i will never grow above the survival threshold β and the alien population fails to invade the i-th site.

Taking into account (2.61), from (2.62) we obtain that propagation of invasive waves become impossible for the following values of the inter-site coupling:

$$D < \frac{1}{K} \ |\min F(U)| \ . \tag{2.63}$$

Note that, contrary to the space-continuous case where wave propagation/blocking depends on integral properties of the population growth rate, in the space-discrete case the condition of propagation failure depends on the details of the growth rate density-dependence in the intermediate range of population density.

A relation similar to (2.63) can be obtained as well for blocking of retreat population waves. In this case, the species retreats from the i-th site if, as a result of the interplay between the local population growth and the inter-site migrations, it appears that $dU_i/dt < 0$. Thus, propagation of retreat waves becomes impossible under the following sufficient condition:

$$-J_{max} + \max F(U) > 0 \tag{2.64}$$

that is, for

$$D < \frac{1}{K} \ \max F(U) \ , \tag{2.65}$$

where the maximum value is taken over the range where the population growth rate is positive, i.e., for $\beta < N/K < 1$.

For parameter values satisfying both (2.63) and (2.65), propagation of neither invasive waves nor retreating waves is possible; thus, it corresponds to a stationary population distribution over space. For instance, such a stationary distribution may have the form of a standing interface separating invaded and non-invaded areas.

For any specific parameterization of $F(U)$, the critical relation between D and β can be obtained explicitly. As an example, let us consider the piecewise linear approximation:

$$F(U) = -\alpha U \quad \text{for} \quad 0 \le U/K \le \beta \ ,$$
$$F(U) = \alpha(K - U) \quad \text{for} \quad U/K > \beta \ ,$$

so that

$$\min F(U) = -\alpha\beta K \quad \text{and} \quad \max F(U) = \alpha(1-\beta)K \ . \tag{2.66}$$

It is readily seen that, in the space-continuous system, the condition of propagation failure $M = 0$ corresponds to $\beta = 0.5$ so that no invasion wave is possible for $\beta \ge 0.5$ and no retreat wave is possible for $\beta \le 0.5$. In the corresponding space-discrete system, however, making use of relations (2.63) and (2.65) we obtain that invasion wave is blocked for

$$D < \alpha\beta \ , \quad 0 < \beta \le 0.5 \tag{2.67}$$

and retreat wave is blocked for

$$D < \alpha(1-\beta) \ , \quad 0.5 \le \beta < 1 \ . \tag{2.68}$$

Relations (2.63) and (2.65) can be combined into a single condition of wave blocking:

$$\frac{D}{\alpha} \ < \ \beta \ < \ 1 - \frac{D}{\alpha} \ . \tag{2.69}$$

Note that inequalities (2.63) and (2.65) or (2.69) provide only a sufficient condition of propagation failure which is the more exact the smaller D is. In the case of piecewise linear growth function, a rigorous analysis appears to be possible that lead to the following exact result (Fath, 1998): wave propagation is blocked if

$$\beta_- \ < \ \beta \ < \ \beta_+ \tag{2.70}$$

where

$$\beta_\pm = \frac{1}{2}\left[1 \pm \left(1 + \frac{4D}{\alpha}\right)^{-1/2}\right] . \tag{2.71}$$

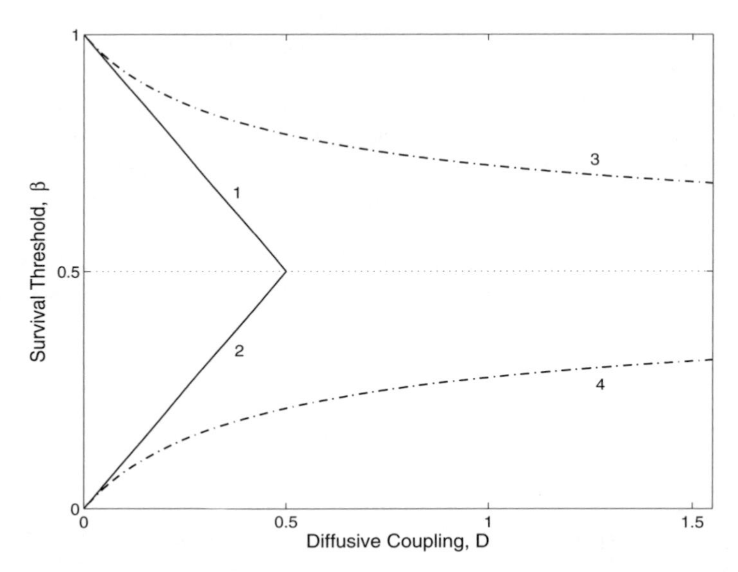

FIGURE 2.4: The structure of the parameter plane for the single-species space-discrete model. Here dashed-and-dotted curves 3 and 4 show the boundary of the domain where parameters correspond to invasion blocking; solid lines 1 and 2 show the sufficient condition of invasion blocking obtained from simple heuristic arguments.

Conditions (2.69) and (2.70–2.71) are shown in Fig. 2.4. Thus, instead of the single value $\beta = 0.5$, we now have a range of values (between curves 3 and 4) where wave propagation is blocked.

Note that the symmetric shape of the invasion pinning domain in Fig. 2.4 is a consequence of symmetric parameterization of the growth rate: it is readily seen that the piecewise-linear function $F(U)$ satisfies the following condition:

$$F(U, \beta) = \; - F(1 - U, 1 - \beta) \; . \tag{2.72}$$

In an arbitrary case, the symmetry may be absent.

In conclusion, it should be mentioned that the discreteness of the system manifests itself also for large values of inter-site coupling when the system properties are intuitively expected to be close to the properties of the continuous system. Considering the large D limit, Keener (1987) showed that the speed of the wave is given by the following equation:

$$c = c_0 (D - D_*)^{1/2} \tag{2.73}$$

where coefficient c_0 is the same as in the equation for the wave speed in the corresponding space-continuous system and D_* is a certain constant. This result has a clear biological meaning: environment fragmentation always leads to lower rates of species invasion even in the case when each habitat is equally favorable for population functioning and the inter-habitat coupling is strong.

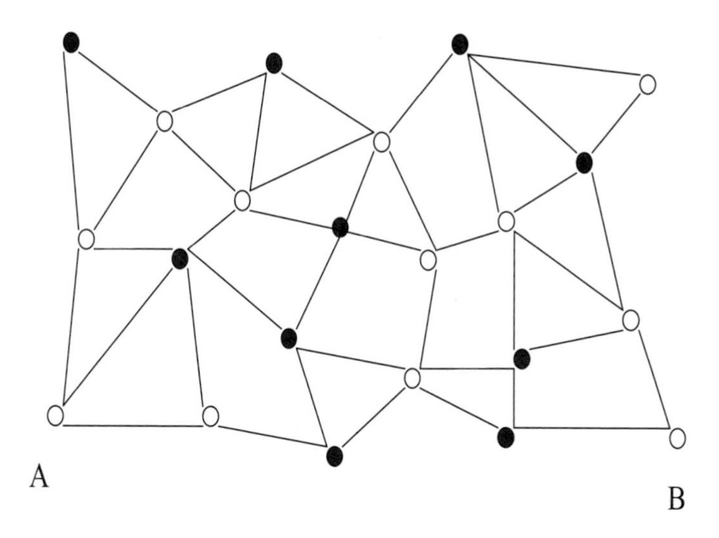

FIGURE 2.5: A sketch of an invasive population spreading through a two-dimensional array of coupled discrete habitats. Here white spots and black spots show "good" and "bad" habitats, respectively. Invasive species can spread from site A to site B only in case there is a chain of "good" habitats connecting A and B.

The above considerations were concerned with the 1-D case. In real ecosystems, the spread of invasive species more often takes place in two spatial dimensions. Assuming that the invasion goes on in a fragmented environment, the continuous space can be reduced to an ensemble of sites or habitats coupled through species migration; see Fig. 2.5. Although this system does not include space explicitly, some features of the sites' relative position and/or the properties of the "borders" separating them can be taken into account by means of inclusion/exclusion of corresponding inter-site links. As a result, contrary to 1-D case, the configuration of the system may become very complicated.

It hardly makes sense to talk about traveling waves in such a system, especially when the system has an irregular structure. Apparently, the patterns of species spread in a 2-D space-discrete system are much more complicated. It seems, however, that each event of the local invasion, i.e., from an i-th site to the neighboring site(s), should follow the scenario qualitatively similar to that considered above.

Let us consider the case when all the habitats are identical in their "quality," i.e., the population dynamics in each of them is described by the same function $F(U)$, and the intensity of diffusive coupling is the same for all pairs, $D_{ij} = D$ for any $i, j = 1, \ldots, N$. Consider the situation when the invasive species has already occupied the i-th site but its neighbors remain virtually empty.

Migration of species is going from site i simultaneously to all its neighbors $1, 2, \ldots, n_i$; thus, the corresponding population flux is split to n_i equal parts so that instead of Eq. (2.61), we now have

$$J_{max} = \frac{1}{n_i} DK . \tag{2.74}$$

Taking into account (2.62), an approximate condition that the invasive species is blocked at the i-th site with n_i neighbors is as follows:

$$\frac{1}{n_i} DK + \min F(U) < 0 . \tag{2.75}$$

Condition (2.75) can be readily generalized to the case when not all of the neighboring sites are empty. Let m_i be the number of the i-th site neighbors that are invaded as well. Then the number of sites actually receiving the population flux from the i-th site is $(n_i - m_i)$ and, instead of (2.75), we arrive at

$$\frac{1}{n_i - m_i} DK + \min F(U) < 0 . \tag{2.76}$$

Note that in the case that the empty sites around the i-th habitat are not connected to any other invaded site, inequality (2.75) provides a sufficient condition of invasion blocking.

Another ecologically relevant problem arises in the case when the habitats are not of equal "quality." In practice, some of them are favorable for population functioning, and thus can be invaded, while others are unfavorable. We assume that unfavorable habitats are unfavorable enough so that they cannot be invaded and remain empty; see Fig. 2.5. Thus, the invasive species can spread through the favorite sites but cannot spread through the unfavorable ones. A question of ecological importance is for what configuration(s) of the system the alien species is blocked at some point and thus will not go through the system from the site(s) of its original introduction. For a system with a large number of sites when a statistical description is justified, this question can be also re-formulated as for what concentration p of the "bad" sites the alien species is going to be blocked and will not invade all favorable habitats.

Apparently, this problem is formally similar to the percolation problem and thus some methods developed in the percolation theory can be applied, e.g., see Stauffer and Aharony (1992). It appears that, under some rather general assumption, there exists a critical value of the bad sites' concentration p_c so that for $p < p_c$ the alien species invades the whole domain while for $p > p_c$ it is trapped inside a certain area. The exact value of p_c depends on the system geometry; in the case of a triangulated domain (when each site has exactly three neighbors), $p_c = 0.5$. Although direct ecological applications of this approach are somewhat impeded by configuration-dependence of the critical concentration, it can still be very helpful in understanding invasive species dynamics under variable environmental conditions; examples can be found in Petrovskii (1998) and Sander et al. (2002).

2.4 Stochastic models

It is a widely recognized fact that the dynamics of ecological populations is affected by numerous factors of a different nature. While some of the factors or feedbacks can be considered as deterministic, such as predator population response to prey density or algae response to sunlight intensity, others are more likely to be stochastic, e.g., those related to fluctuations in the weather conditions. Relative importance of deterministic and stochastic factors for population dynamics is often not clear and the issue as a whole has been a subject of intense discussions.

An observation that makes this issue even more controversial is that sometimes mathematical models based on deterministic and stochastic approaches lead to very similar results; one example can be found in recent works by Kawasaki et al. (1997) and Mimura et al. (2000). Moreover, in many cases the models based on deterministic equations are successfully used to describe the processes of a clearly stochastic nature; a classical example is given by the diffusion equation. Although application of the diffusion equation is somewhat restricted by the underlying assumption that the density of diffusing particles/individuals should be sufficiently high (in order to keep fluctuations small), it was shown by Skellam (1951) that a random motion of a single individual can be also described by the diffusion equation as soon as we treat its solution as the probability distribution function to locate the individual around a given position in space.

Nevertheless, for the level of "intermediate complexity," i.e., when there is more than one individual but the population density is still low, fluctuations in population density can be important and thus the diffusion approximation may be not good enough. One rigorous theoretical approach applicable to this case is based on multi-particle probability distribution functions and the master equation. There is extensive literature concerned with this issue, e.g., see Horsthemke and Lefever (1984), Allen (2003) and references therein; for the sake of brevity we do not go into details of this approach here.

For the situations when the probabilistic nature of population dynamics is important, there is also a more straightforward modeling approach based on cellular automata. According to this approach, the continuous space is mimicked by a space-discrete ensemble of sites, or "cells." The cells are numbered and each cell has an associated variable or variables. Let the state of the i-th cell be described by $a_i(t)$ which may have different biological meaning depending on the system being modelled and the purposes of modeling. For instance, it can be the number of the individuals of a given species occupying the site; alternatively, in case one site cannot be occupied by more than one individual, it can describe the state of the individual, e.g., its age. Starting from an initial condition, $a_i(0) = a_{i0}$, $i = 1, \ldots, N$ where N is the total number of cells in the system, the state of the cells throughout the system is then updated simultaneously at discrete moments $t_1, t_2, \ldots$ according certain prescribed rules. Applications of this approach to real-world ecological problems can be found in Higgins et al. (1996), Higgins and Richardson (1996) and Cannas et al. (2003).

As an example, let us consider in some more details how the cellular automata model is applied to study an invasion of alien plant species. In this case, it is reasonable to assume that a given site or cell cannot be occupied by more than one individual (which apparently corresponds to the assumption that the size of the sites is small enough). The state variable a_i can then conveniently be treated as the age of the individual in the i-th cell; $a_i = 0$ in case the cell is empty. The interval $\Delta t = t_{k+1} - t_k$ between the two consequent moments when the cell variables are updated corresponds to the minimum reproductive interval in the species' life history, e.g., $\Delta t = 1$ year. An occupied cell is updated according to the following rule:

$$a_i(t_{k+1}) = a_i(t_k) + 1 \text{ with probability } q \ , \tag{2.77}$$

$$a_i(t_{k+1}) = 0 \text{ with probability } 1 - q \tag{2.78}$$

where q can be referred to as the annual survival probability and the value of q is chosen based on biological reasons.

A site i that is empty at the moment t_{k-1} will be colonized at t_k with a certain colonization probability so that $a_i(t_k) = 1$ with probability $p_i(t_k)$ provided that $a_i(t_{k-1}) = 0$. Here, in case of a plant invasion, the colonization probability $p_i(t_k)$ depends on the number of seeds S_i received by the site i at the time t_k. To calculate the number of seeds S_i, one should count the seeds received by the i-th cell from the rest of the cells in the system. Assuming that speed of dispersion is spatially isotropic, the distribution generated by a single individual is described by the density distribution function $f(r)$ where $r = \sqrt{x^2 + y^2}$ is the distance from a parent plant. The fraction of seeds G_{ij} received by the cell i coming from the individual located in the cell j is given by

$$G_{ij} \approx \Omega_i f(r_{ij}) \tag{2.79}$$

where Ω_i is the area of the i-th site and r_{ij} is the distance from the center of cell i to the center of cell j, so that

$$S_i = \sum_{j \neq i} G_{ij} \approx \Omega_i \left(\sum_{j \neq i} f(r_{ij}) \right) \tag{2.80}$$

where the sum runs over all non-empty cells in the system.

Eqs. (2.77–2.80) are then used in computer simulations in order to study invasion patterns subject to parameter values.

The above approach allows for the cases where the origin of stochasticity is in population fluctuations that may be caused, for instance, by low population density. Alternatively, stochasticity in population dynamics can appear as a result of environmental fluctuations or external forcing when certain parameters, e.g., population growth rate and/or mortality, vary in space and time in a stochastic manner. Such variation can be regarded as a "noise" and, after making necessary assumptions about the noise properties, can be incorporated into the model straightforwardly, cf. Steele and Henderson (1992b).

For instance, assuming that the species mortality is affected by white multiplicative noise, for a rather general model of population dynamics we obtain:

$$\frac{\partial u_i(\mathbf{r},t)}{\partial t} = \nabla\left(D_i \nabla u_i\right) + f_i\left(u_1,\ldots,u_n\right) - \mu_i u_i \xi_i(\mathbf{r},t), \qquad i = 1,\ldots,n \tag{2.81}$$

where μ_i is the average mortality rate of the i-th species and n is the number of species in the model. Here $\xi_i(\mathbf{r},t)$ is a spatiotemporal white Gaussian noise, i.e., a random Gaussian field with zero mean and delta correlation:

$$\langle \xi_i(\mathbf{r},t)\rangle = 0, \ \ \langle \xi_i(\mathbf{r}_1,t_1)\,\xi_i(\mathbf{r}_2,t_2)\rangle = \delta(\mathbf{r}_1-\mathbf{r}_2)\,\delta(t_1-t_2), \qquad i = 1,\ldots,n\ . \tag{2.82}$$

The model (2.81–2.82) has been used to study the interplay between deterministic and stochastic factors in population dynamics. This interplay appears to be highly nontrivial. It was recently shown by Malchow et al. (2002) that there exists a critical level of noise in this system so that the system dynamics is more driven by deterministic mechanisms when noise intensity is small but becomes apparently stochastic for stronger noise, the change between different types of dynamics taking place in a narrow transition region. Even in the supercritical case, when noise can change the system's spatiotemporal dynamics significantly, it was shown that intrinsic spatial scales of the system are still controlled by deterministic mechanisms (Malchow et al., 2004).

The above results make it possible to distinguish between the cases when the rate of spread of an invasive species is likely to be affected by environmental noise and when the impact of stochastic factors is unlikely to be significant. In the case that the noise intensity is not very high, the species invasion can be expected to follow the pattern typical for deterministic systems, i.e., it spreads over space through propagation of population front. The impact of stochasticity is the more substantial the lower is the population density. The population density of alien species is the lowest at the leading edge of the front. In the absence of the Allee effect, the speed of the front is controlled by the population dynamics at the leading edge (cf. Section 7.1), i.e., exactly where stochasticity is expected to be important and thus the speed is likely to be modified by the impact of noise. On the contrary, in the case when population growth is damped by the Allee effect, the speed of invasion is determined by the population dynamics behind the front, i.e., for intermediate and large values of the population density, and thus it is unlikely to be affected by stochastic factors.

2.5 Concluding remarks

In this chapter, we have made a brief excursion into mathematical approaches and methods that are often used for modeling biological invasions. By no means is our review exhaustive or complete. Complexity of ecosystem dynamics together with progress in applied mathematics and computer technologies have brought to life a great variety of mathematical tools that are used in contemporary mathematical ecology. Since this book is primarily concerned with exact solutions of relevant models (which in most cases are based on partial differential equations), we could hardly do more than to just outline a few typical approaches in an attempt to provide some basic ideas how different manifestations of ecosystem complexity, e.g., continuity/discreteness or determinism/stochasticity, can be possibly mimicked by mathematical models. A somewhat wider view on modeling biological invasion can be obtained from Petrovskii and Malchow (2005).

It should be mentioned that, although the large number of available mathematical tools is certainly a positive factor enhancing theoretical studies, a question may arise about consistency of different approaches. It often happens that the mathematical model of given ecological phenomenon is chosen based not on the specifics of the problem but more on personal preference or personal experience of the researcher(s) doing the research. In general, however, this situation should not be necessarily regarded as negative: indeed, what bad can be in people's intention to use their qualification and experience in order to accomplish the study more effectively? Different approaches often lead to qualitatively similar results – as far as they are formulated and/or used properly. For instance, it may seem at first glance that the space-continuous models considered in Section 2.1 are not capable to catch the phenomenon of wave blocking due to environment fragmentation described by the space-discrete models, cf. equations (2.22) and (2.70–2.71), also (2.20) and (2.73). This discrepancy, however, can be resolved immediately in case a sort of spatial heterogeneity is incorporated into the diffusion-reaction models through explicit space-dependence of corresponding parameters. The modified models then exhibit properties qualitatively similar to those described by the space-discrete approach; in particular, wave blocking and wave propagation slow-down have been observed (Shigesada et al., 1986; Barenblatt et al., 1995; Petrovskii, 1997, 1998; Kinezaki et al., 2003).

Similar arguments apply to the relation between the (advection-)diffusion-reaction models and integral-difference models. Although the latter certainly provides a more general description of species dispersal, their power seems to be exaggerated. In particular, accelerating population waves of invasive species can be described as well by diffusion-reaction models in case the scale-dependence of the diffusion coefficient is taken into account (Petrovskii, 1999b). Such scale-dependence was shown to be a common property of the turbulence-driven dispersal (Okubo, 1980), no matter whether the species is water-borne or air-borne.

The situation is somewhat more complicated when it concerns the duality of ecosystem dynamics stemming from the interplay between deterministic and stochastic factors. Although this problem is largely open, it seems that a lot of confusion come from a simple misunderstanding of the origin of the "deterministic" models. Ecologists are sometimes too much impressed by seeing irregular fluctuations in ecological data. Indeed, ecosystem dynamics is intrinsically stochastic; however, it is very well known that stochastic processes are described by deterministic equations which are either integral equations or, under certain additional assumptions, partial differential equations (Kolmogorov, 1931; also see Feller, 1971). Thus, it is, in fact, incorrect to treat the models based on diffusion-reaction equations as "purely deterministic." They provide a mean-field description of stochastic processes; what they actually do not take into account is fluctuations in population density. (It should be mentioned that the magnitude of stochastic fluctuations can still be described by a deterministic "fluctuation-dissipation relation," e.g., see Haken (1983).) That brings forward a more general issue about the limits of predictability in ecosystem dynamics. It is a basic principle that it is impossible to predict the exact time and magnitude of a particular stochastic fluctuation (as well as it is impossible to predict the exact position of a quantum particle) and the idea to reproduce every particular hump in given data by means of simulations is weird. From this standpoint, the deterministic models provide us with nearly as much predictive power as we can possibly have.

It should also be mentioned that deterministic models are not exhausted by the advection-diffusion-reaction equations and integral-difference equations. Diffusion approximation is essentially based on the assumption of individual's random walk; in case this assumption is relaxed, another type of PDE-based model may appear. In particular, Holmes (1993) showed that, in the case of a correlated random walk, population dynamics is described by the so-called telegraph equations that predict much higher invasion rates compared to diffusion-reaction models.

Besides the model capacity to describe a given ecological phenomenon, another important feature is its solvability. Accounting for the fact that population dynamics is usually nonlinear, a great majority of mathematical models can only be used by means of computer simulations because existing analytical approaches are often inadequate or insufficient. Since numerical study implies particular parameter values, lack of analytical methods and relevant analytical solutions decrease generality of results and reduce their reliability. Ideally, a theoretical study should combine numerical and analytical approaches, e.g., to be based on a mathematical model that can be solved analytically for a certain special case. It has so happened that most of the analytical solutions have been obtained for diffusion-reaction models. Those of them that are likely to have application to modeling biological invasion will be revisited in the next chapters.

Chapter 3

Basic methods and relevant examples

In this chapter, we give a review of several methods and approaches often used to construct exact solutions of nonlinear partial differential equations, in particular, diffusion-reaction equations. Since a general theory is missing in most cases, and its development certainly lies beyond the scope of this book, it is more instructive to revisit them by means of considering how they can be applied to particular cases rather than by giving a formal description. Also, as it fully complies with the purposes of this book, we focus on the properties of the obtained exact solutions. A question of particular interest is how the speed of the traveling fronts depends on the parameter values.

The first of the revisited approaches, see Sections 3.1 and 3.2, is based on the idea of Hopf (1950) that a successful change of variables can linearize the given nonlinear equation. This method, although powerful in some cases, cf. "C-integrable equations" (Calogero and Xioda, 1991), essentially depends on the choice of the substitution form. We will consider a certain generalization to the classical Cole–Hopf transformation and show how it works when applied to diffusion-reaction equations.

Section 3.3 deals with the approach based on a direct linearization of the nonlinearities contained in the equation(s) under study. In the case of diffusion-reaction models, the source of nonlinearity is usually found in the density-dependence of the population growth rate. The direct linearization method assumes that the growth function can be approximated by a few straight lines so that the original equation breaks into a system of linear PDEs.

Finally, we examine how nonlinear partial differential equations can possibly be solved without linearization; see Sections 3.4 and 3.5. One typical approach is based on using a relevant ansatz, i.e., a prescribed solution structure. This method requires preliminary information regarding the form of the ansatz. That can often be found in intrinsic symmetries of given equations. We will consider a few instructive examples of this methods application and try to give some ideas of how the solution structure can be foreseen.

3.1 The Cole–Hopf transformation and the Burgers equation as a paradigm

The Burgers equation provides a simple model combining linear diffusion with nonlinear transport:

$$u_t - 2Auu_x = Du_{xx} \tag{3.1}$$

where $u = u(x,t)$ is the state variable, x is the position in space, t is time, A and D are parameters and the coefficient (-2) on the left-hand side is introduced for convenience. In order to avoid ambiguousness, we assume that $A > 0$; clearly, the case $A < 0$ corresponds to the change in the direction of axis x.

Equation (3.1) is one of the most famous equations in nonlinear sciences. Initially proposed as a model of turbulent flow (Burgers, 1948), it was later applied to many other processes of different origin as well. In particular, it has recently been applied to describe species dispersal in population dynamics (Berezovskaya and Khlebopros, 1996; Berezovskaya and Karev, 1999) with the right-hand side describing, as usual, the individual random motion and the second term on the left-hand side accounting for density-dependent migrations; see also Section 4.1. It should be mentioned, however, that since Eq. (3.1) does not contain the "reaction" terms accounting for the local processes such as birth, death, etc., its possible ecological applications are limited to situations when population multiplication can be neglected.

There is extensive literature concerned with the Burgers equation and its solutions; probably the most exhaustive source is Sachdev (1987). In this section, we make only a brief review of its properties to the extent that makes it useful for our subsequent analysis.

For studying general mathematical properties of Eq. (3.1), the coefficients A and D are not important. By means of introducing dimensionless variables,

$$\tilde{u} = \frac{u}{U_0}, \quad \tilde{t} = \left(\frac{A^2U_0^2}{D}\right)t, \quad \tilde{x} = \left(\frac{AU_0}{D}\right)x,$$

where U_0 is a certain characteristic value to be taken, for instance, from the initial conditions $u(x,0) = u_0(x)$, from Eq. (3.4) we arrive at

$$u_t - 2uu_x = u_{xx} \tag{3.2}$$

(omitting tildes for notation simplicity).

In the early 50's of the last century, Hopf (1950) and Cole (1951) independently showed that the Burgers equation is equivalent to the linear diffusion equation for a new variable $U(x,t)$,

$$U_t = U_{xx} \ , \tag{3.3}$$

by means of the following transformation:

$$u(x,t) = \frac{U_x}{U} \tag{3.4}$$

so that, for any solution U of equation (3.3), u given by (3.4) is a solution of (3.2).

Indeed, substituting (3.4) into (3.2), we obtain

$$\left(\frac{U_{xt}}{U} - \frac{U_x U_t}{U^2}\right) - 2\frac{U_x}{U}\left(\frac{U_{xx}}{U} - \frac{U_x^2}{U^2}\right) \tag{3.5}$$
$$= \left(\frac{U_{xxx}}{U} - 3\frac{U_{xx}U_x}{U^2} + 2\frac{U_x^3}{U^3}\right)$$

so that, after simple algebra, we arrive at

$$\frac{(U_t - U_{xx})_x}{U} - \frac{U_x(U_t - U_{xx})}{U^2} = 0 \tag{3.6}$$

which, assuming that $U > 0$ for all x and t, is equivalent to (3.3).

Clearly, the reverse transformation is also possible:

$$U(x,t) = \exp\left(\int u(x,t)dx\right). \tag{3.7}$$

Relations (3.4) and (3.7) make it possible to write the *general solution* of nonlinear equation (3.2) explicitly. Let us consider the initial-value problem for Eq. (3.2) in an infinite domain, $-\infty < x < \infty$, with the initial conditions being described by a certain function $u_0(x)$. By virtue of (3.7), the corresponding initial conditions for Eq. (3.3) are

$$U(x,0) = \Phi(x) = \exp\left(\int_{-\infty}^{x} u_0(\zeta)d\zeta\right). \tag{3.8}$$

The solution of the diffusion equation with the initial condition (3.8) is then given as

$$U(x,t) = \frac{1}{\sqrt{4\pi t}}\int_{-\infty}^{\infty} \exp\left(-\frac{(x-y)^2}{4t}\right)\Phi(y)dy\,, \tag{3.9}$$

cf. Appendix. Eqs. (3.4) and (3.9) provide the analytical solution of (3.2).

While the general solution given by (3.4), (3.8) and (3.9) describes the evolution of the initial conditions, in the large-time asymptotics the Burgers equation has solutions corresponding to traveling wave fronts. This is one of its properties that make it relevant (to a certain extent, cf. the comments at the beginning of this section) to biological invasion modeling.

The existence of the traveling fronts can be shown straightforwardly. Let us consider the following conditions at infinity:

$$u(x \to -\infty, t) = u_- \ , \quad u(x \to \infty, t) = u_+ \ . \tag{3.10}$$

We look for a traveling wave solution of the Burgers equation, i.e., $u(x,t) = v(\xi)$ where $\xi = x - ct$, c being the speed of the wave. Eq. (3.2) takes the form

$$\frac{d^2v}{d\xi^2} + 2v\frac{dv}{d\xi} + c\frac{dv}{d\xi} = 0 \ . \tag{3.11}$$

Let $dv/d\xi = \phi(v)$ where $\phi(v)$ is an unknown function to be determined. Since

$$\frac{d^2v}{d\xi^2} \ = \ \frac{d}{d\xi}\phi(v) \ = \ \frac{d\phi}{dv}\frac{dv}{d\xi} \ = \ \phi\frac{d\phi}{dv} \ , \tag{3.12}$$

from (3.11), we obtain:

$$\frac{d\phi}{dv} + 2v + c = 0 \tag{3.13}$$

(assuming that $\phi > 0$) so that

$$\phi(v) = \ - v^2 - cv + C_0 \tag{3.14}$$

where C_0 is the integration constant.

Now we can make use of the conditions at infinity. Evidently, (3.10) imply that

$$\frac{dv(\xi \to \pm\infty)}{d\xi} = 0 \tag{3.15}$$

which means that $\phi(u_-) = \phi(u_+) = 0$. Thus, u_- and u_+ are the roots of the square polynomial (3.14) so that

$$\phi(v) = \ - (v - u_-)(v - u_+) \ . \tag{3.16}$$

From comparison between (3.14) and (3.16), we obtain that $C_0 = -u_-u_+$. Moreover, we also obtain the equation for the wave speed:

$$c = \ - (u_- + u_+) \tag{3.17}$$

or, in the original dimensional variables,

$$c = \ - (u_- + u_+) \, AU_0 \ . \tag{3.18}$$

Recalling that $\phi = dv/d\xi$, Eq. (3.16) is readily solved:

$$\log\left|\frac{v - u_+}{v - u_-}\right| = (u_- - u_+)(\xi - \xi_0) \ , \tag{3.19}$$

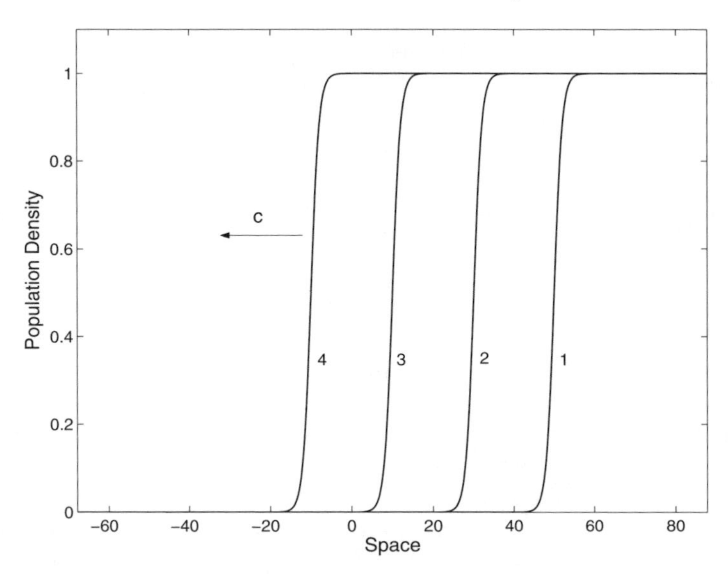

FIGURE 3.1: Propagation of traveling population wave as given by the exact solution (3.21) of the Burgers equation. Curves 1 to 4 show the wave profiles at $t = 0,\ t = 20,\ t = 40$ and $t = 60$, respectively.

and finally we arrive at the following traveling wave solution:

$$u(x,t) \;=\; v(\xi) = \frac{u_+ + u_- \exp[(u_- - u_+)(\xi - \xi_0)]}{1 + \exp[(u_- - u_+)(\xi - \xi_0)]} \;, \tag{3.20}$$

where the integration constant ξ_0 quantifies the front position at $t = 0$.

Keeping in mind possible application of (3.20) to biological invasions, we consider $u_- = 0,\ u_+ = 1$. In this case, (3.20) takes a somewhat simpler form:

$$u(x,t) \;=\; v(\xi) = \left(1 + \exp^{-(\xi - \xi_0)}\right)^{-1} \;. \tag{3.21}$$

The corresponding value of the wave speed is $c = -1$.

Solution (3.21) is shown in Fig. 3.1. Here the direction of the front propagation (shown by the arrow) is determined by the sign of coefficient A in the original equation (3.1); indeed, the factor $-2Au$ corresponds to the wave speed A_0 in Eq. (1.21) describing propagation of the simple wave. This can also be seen from the formal derivation of the solution (3.20). Note that the conditions at infinity affect the speed value but not its sign. This kind of dependence of the wave speed on the problem parameters is essentially different from that observed for diffusion-reaction equations, cf. Section 2.1. In the latter case, it is the direction of the wave propagation that is determined by the conditions at infinity while its value is determined by the equation coefficients, e.g., see Eq. (2.19). An interesting question thus arises about a possible outcome of the interplay between these two different mechanisms of wave propagation. This problem will be considered in Section 4.1.

3.1.1 * Exact solutions for a forced Burgers equation

The Burgers equation with a nontrivial right-hand side, that is

$$u_t - 2uu_x - u_{xx} = F \tag{3.22}$$

where F is an external "force," is usually called the forced Burgers equation. In case Eq. (3.22) is applied to population dynamics, F is likely to depend on u in a manner consistent with the features of the local population growth, cf. Section 1.2, and may also depend on x and t as a result of, for instance, environmental heterogeneity and transient weather conditions. The general case $F = F(x,t,u)$ is, however, very difficult to treat analytically and it is unlikely that any solution can be obtained a closed form. Instead, we will consider separately two cases, i.e., $F = F(u)$ and $F = F(x,t)$. In this section, we will focus on the latter case; the case $F = F(u)$ will be studied in detail in Section 4.1. It must be mentioned that the biological meaning of Eq. (3.22) with density-independent forcing is somewhat obscure and the contents of this section should be regarded more as an example of linearization technique rather than a model of immediate biological relevance.

It is readily seen that the Cole–Hopf substitution (3.4) transforms the forced Burgers equation into a linear equation for the new variable $U(x,t)$:

$$U_t - U_{xx} = U \int_{x_0}^{x} F(\zeta, t) d\zeta \; . \tag{3.23}$$

In spite of the fact that Eq. (3.23) is linear, constructing its analytical solution for an arbitrary function $F(x,t)$ is a difficult problem. A general algorithm for constructing a solution in the form of a series under some nonrestrictive assumptions regarding $F(x,t)$ was earlier developed by O.A.Oleinik with collaborators. However, the expressions appearing as a result of their method are so cumbersome that they are of little practical use because it is hardly possible to investigate how the solution properties change with parameter values. For that reason, instead of studying Eq. (3.23), we apply another approach (Petrovskii, 1999a).

Stationary forcing. We begin with the case when the "force" F is stationary, i.e., does not depend on time. Instead of the classical substitution (3.4), we consider its modification:

$$u = \frac{U_x}{U} + k(x) \; , \tag{3.24}$$

where $k(x)$ is a certain function to be defined.

Substituting (3.24) into (3.22), after some standard transformations we obtain:

$$\left(\frac{U_t - 2kU_x - U_{xx}}{U}\right)_x = \left(\frac{dk}{dx} + k^2 + F(x) - C(t)\right)_x$$

where $C(t)$ is an arbitrary function of time (minus is chosen for convenience).

Further on, integrating over x we obtain:

$$U_t - 2kU_x - U_{xx} = \left(\frac{dk}{dx} + k^2 + \psi(x) - C(t)\right) U$$

where ψ is the antiderivative of F, i.e., $d\psi/dx = F(x)$.

Thus, if $k(x)$ is a solution of the Riccati equation

$$\frac{dk}{dx} + k^2 = -\psi(x) + C , \tag{3.25}$$

then $U(x,t)$ is a solution of the linear advection-diffusion equation:

$$U_t - 2kU_x = U_{xx} . \tag{3.26}$$

Here Eq. (3.25) may contain time only as a parameter, i.e., as the argument of function C. Let us assume that $C(t) = const$. Then Eq. (3.25) coincides with the forced Burgers equation integrated over space in the stationary case $u_t \equiv 0$. Thus, k is a stationary solution of (3.22) and the meaning of substitution (3.24) becomes clear: Eq. (3.26) for the new state variable describes a dynamical process going "on top" of the stationary background density $k(x)$.

To solve the Riccati equation with the right-hand side of an arbitrary form is not an easy problem and it does not always appear possible to construct a solution in a closed form. Still, in some cases system (3.25–3.26) turns out to be more convenient for obtaining biologically/physically meaningful exact solutions of the forced Burgers equation than the approach based on Eq. (3.23).

Let us try to look for a traveling wave solution of Eq. (3.26), i.e., assuming $U(x,t) = v(\xi)$ where $\xi = x - y(t)$. Having substituted it into (3.26), we obtain:

$$-\left(\frac{dy}{dt} + 2k(x)\right)\frac{dv}{d\xi} = \frac{d^2v}{d\xi^2} . \tag{3.27}$$

The transition to traveling wave coordinates is mathematically correct only in the case that Eq. (3.27) contains x and t through the new variable ξ. It immediately implies that $dy/dt + 2k(x) = \phi(\xi)$ where ϕ is a certain function. Since x and ξ are related through a linear equation, see the definition of ξ, this is only possible when k and ϕ are linear functions, i.e., $k(x) = Bx + B_1$ and $\phi(\xi) = \beta\xi + \gamma$ where B, B_1, β, γ are parameters. For the linear $k(x)$, Eq. (3.25) is reduced to

$$B + (Bx + B_1)^2 = -\psi(x) + C$$

and, differentiating it with respect to x, we obtain that $F(x) = -2B(Bx + B_1)$. Note that we can let $B_1 = 0$ without loss of generality by means of an appropriate choice of coordinates, $x \to x - B_1/B$. Thus, the forced Burgers equation has a traveling wave solution only if the forcing is linear with respect to x, i.e., $F(x) = -2B^2x$. Correspondingly, $C = B$ and $k(x) = Bx$.

From Eq. (3.27), we obtain:

$$\frac{dy}{dt} + 2Bx = \beta\xi + \gamma \quad = \quad \beta(x - y(t)) + \gamma$$

which can be written as

$$\frac{dy}{dt} + \beta y - \gamma = (\beta - 2B)x \ .$$

The last equation only makes sense in case both of its sides are equal to zero. Therefore, $\beta = 2B$ and

$$\frac{dy}{dt} + \beta y - \gamma = 0 \ . \tag{3.28}$$

Under the natural assumption that $\xi = x$ for $t = 0$, we obtain that $y(0) = 0$ and thus the solution of Eq. (3.28) is

$$y(t) = \delta\left(1 - e^{-2Bt}\right)$$

where $\delta = \gamma/(2B)$.

Taking into account all that we know now about y and k, from (3.27) we obtain:

$$-\left(2B\xi + \gamma\right)\frac{dv}{d\xi} = \frac{d^2v}{d\xi^2} \ .$$

Introducing $p(\xi) = dv/d\xi$, the above equation is reduced to

$$\frac{dp}{d\xi} = -\left(2B\xi + \gamma\right)p$$

which is readily solved to yield the expression for p:

$$p \equiv \frac{dv}{d\xi} = const \cdot \exp\left(-B\xi^2 - \gamma\xi\right) . \tag{3.29}$$

The case $B < 0$ is not biologically meaningful since in this case $p(\xi)$ would increase unboundedly for $|x| \to \infty$. Therefore, we let $B > 0$. Then the solution of Eq. (3.29) is:

$$v(\xi) = a + b\left(1 + \text{erf}\left[B^{1/2}(\xi + \delta)\right]\right)$$

where a and b are parameters determined by the initial conditions and $\text{erf}(z)$ is the error function.

Coming back to initial variables x, t and U, we obtain:

$$U(x,t) = a + b\left(1 + \operatorname{erf}\left[B^{1/2}\left(x + \delta e^{-2Bt}\right)\right]\right).$$

Finally, taking into account (3.24) we arrive at the following exact solution of the Burgers equation with linear forcing:

$$u(x,t) = Bx + \frac{\exp\left[-B\left(x + \delta e^{-2Bt}\right)^2\right]}{\mu + [\pi/(4B)]^{1/2}\operatorname{erf}\left[B^{1/2}\left(x + \delta e^{-2Bt}\right)\right]} \tag{3.30}$$

where $\mu = [\pi/(4B)]^{1/2}(a+b)/b$. Obviously, for $|\mu| > [\pi/(4B)]^{1/2}$ function $u(x,t)$ given by Eq. (3.30) is continuous for any x and t. It is readily seen that solution (3.30) describes propagation of a dome-shaped asymmetric wave along the background stationary profile $k(x) = Bx$. Here δ and μ are parameters determined by the initial conditions so that δ describes the position of the wave at $t = 0$ and μ describes its amplitude.

Note that, due to the linearity of Eq. (3.26), a linear combination of its solutions is also a solution. Therefore, a more general N-wave solution of the Burgers equation with linear forcing has the following form:

$$u(x,t) = Bx + \frac{\sum_{i=1}^{N}\epsilon_i \exp\left[-B\left(x + \delta_i e^{-2Bt}\right)^2\right]}{\mu + [\pi/(4B)]^{1/2}\sum_{i=1}^{N}\epsilon_i \operatorname{erf}\left[B^{1/2}\left(x + \delta_i e^{-2Bt}\right)\right]}. \tag{3.31}$$

If the constants δ_i differ strongly enough, the solution (3.31) describes a set of N individual waves which gradually approach the origin. An example is shown in Fig. 3.2 where interaction and merging of three traveling humps results, in the large-time limit, in formation of a stationary profile situated around $x = 0$. Thus, Eq. (3.22) with stationary linear forcing describes wave blocking due to the effect of forcing. In the case when constants δ_i are of the same order, the individual waves are not distinguishable and solution (3.31) describes the evolution of initial profile that may have a very complicated form.

Transient forcing. In the case considered above, the right-hand side of the forced Burgers equation depended only on coordinate x. In a more general case, the forcing is not stationary and $F = F(x,t)$. It is straightforward to see that in this case transformation (3.24) also leads to linear equation (3.26) where k now depends on x and t. However, the coupling equation is no longer a Riccati equation but coincides with the original equation (3.22). Thus, in the case of transient forcing, substitution (3.24) describes an "autotransformation" of the solutions so that if $k(x,t)$ is a solution of the forced Burgers equation with an arbitrary F then

$$u(x,t) = \frac{U_x}{U} + k(x,t) \tag{3.32}$$

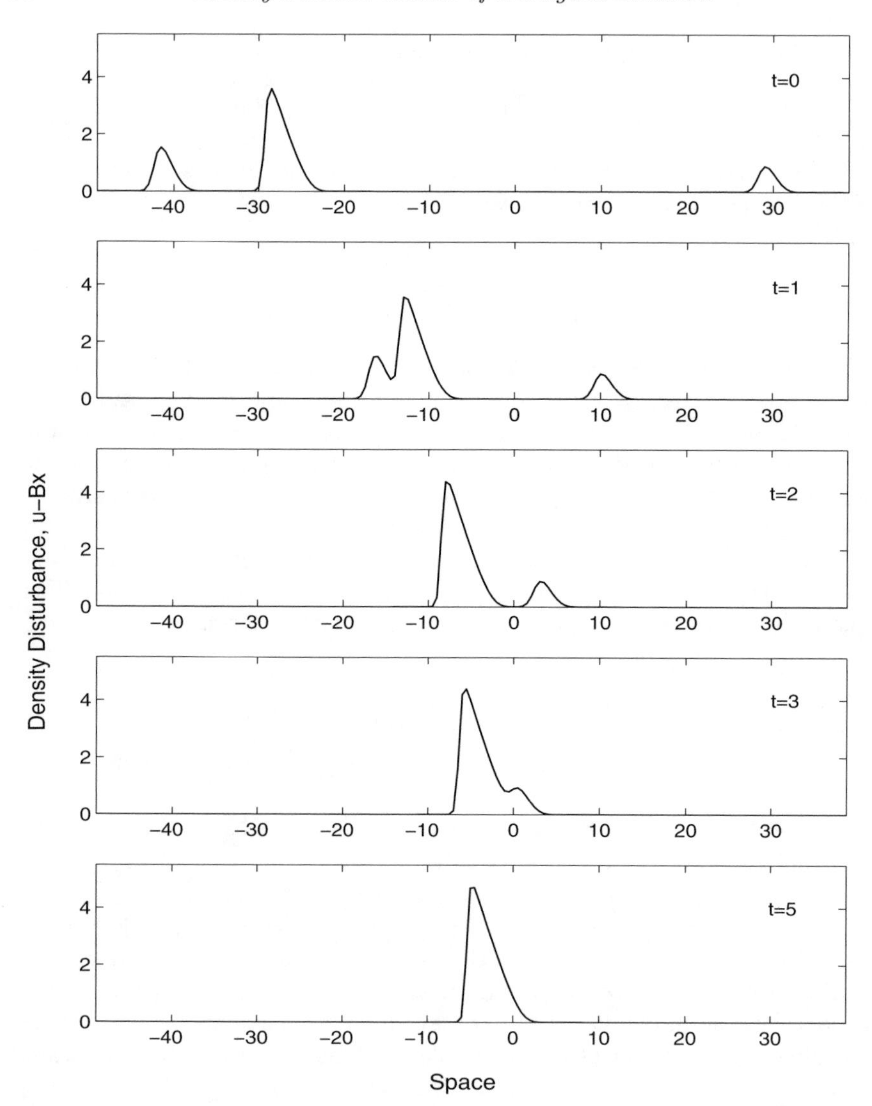

FIGURE 3.2: Evolution of a three-hump disturbance of the stationary profile $k(x) = Bx$ as given by the exact solution (3.31) of the forced Burgers equation. Parameters are: $\delta_1 = 40,\ \delta_2 = 25,\ \delta_3 = -30,\ \epsilon_1 = 0.000001,\ \epsilon_2 = 0.1,\ \epsilon_3 = 0.899999,\ B = 0.5$ and $\mu = 1.2532957$.

is another solution (corresponding to different initial conditions), provided that $U(x,t)$ is a solution of Eq. (3.26).

Relation (3.32) can be used to construct exact solutions of the forced Burgers equation when F depends on time. As an example, let us consider the special case when forcing is decaying with time:

$$u_t - 2uu_x - u_{xx} = -\frac{ax}{(t+t_0)^2} \tag{3.33}$$

where a and t_0 are parameters. In order to avoid singularities for $t > 0$, we assume that $t_0 > 0$.

It is readily seen that the function

$$k(x,t) = \frac{bx}{t+t_0} \tag{3.34}$$

is a solution of Eq. (3.33) if $b + 2b^2 = a$. The solutions of Eq. (3.33) may have different properties depending on the sign of a and b. Since our goal here is more to show how the autotransformation (3.32) can be used to generate exact solutions of the Burgers equation with transient forcing rather than to investigate it in all details, we restrict our consideration to the case $a > 0$.

The solution (3.34) by itself is unlikely to be of much interest because its behavior is too simple. However, it can be used to construct other solutions with more interesting properties. By virtue of (3.32), the function $u = k + (U_x/U)$ is also a solution in case $U(x,t)$ is a solution of the following equation:

$$U_t - \frac{2bx}{t+t_0} U_x = U_{xx} \ . \tag{3.35}$$

The combination of x and t in which they appear in Eq. (3.35) gives us a hint that it may be possible to look for a self-similar solution, i.e., in the form $u(x,t) = v(\theta)$ where $\theta = x\phi(t)$ and functions v and ϕ are to be determined. Having substituted it into Eq. (3.35) we obtain:

$$\left(x\phi^{-2}\,\frac{d\phi}{dt} - \frac{2bx}{(t+t_0)\phi} \right) \frac{dv}{d\theta} = \frac{d^2v}{d\theta^2} \ .$$

The transition to self-similar variables is mathematically correct only in case the expression in parentheses is a function of θ. In order to satisfy this condition, we require that

$$\phi^{-2}\,\frac{d\phi}{dt} = \lambda\phi \ , \qquad \frac{1}{(t+t_0)\phi} = \eta^{-2}\phi \tag{3.36}$$

where λ and η are certain constants.

From Eqs. (3.36), we immediately arrive at $\phi(t) = \eta(t+t_0)^{-1/2}$, $\lambda = -0.5\eta^{-2}$. Letting $\theta(x,0) = x$, we obtain $\eta = t_0^{1/2}$. Eq. (3.36) then takes the following form:

$$-\,2\alpha^2\,\theta\frac{dv}{d\theta} = \frac{d^2v}{d\theta^2} \ ,$$

where $\alpha^2 = (b + 0.25)/t_0$. The last equation is solved easily to yield the following solution:

$$v(\theta) = A_1 \mathrm{erf}(\alpha\theta) + A_2 \tag{3.37}$$

where A_1 and A_2 are parameters determined by the initial conditions. Taking into account (3.32), we arrive at the following exact solution of Eq. (3.33):

$$u(x,t) = \frac{bx}{t+t_0} + \frac{2\alpha}{\sqrt{\pi}}\left(\frac{t_0}{t+t_0}\right)^{1/2} \frac{\exp(-\alpha^2\theta^2)}{\kappa + \mathrm{erf}(\alpha\theta)} \tag{3.38}$$

where $\kappa = (A_2/A_1)$ and $\theta = x[t_0/(t+t_0)]^{1/2}$.

For $|\kappa| > 1$, the function given by (3.38) is continuous at all x and $t > 0$. Exact solution (3.38) describes self-similar diffusion and decay of a dome-shaped initial disturbance of the linear density distribution. Thus, the simple "tentative" solution (3.34) was used, by means of autotransformation (3.32), to generate a more interesting solution (3.38).

3.2 Further application of the Cole–Hopf transformation

In this section, we will consider a certain generalization of the linearization technique provided by the Cole–Hopf transformation (3.4). The focus of our consideration will be on the single-species model of population dynamics allowing for the strong Allee effect, the growth rate being described by a cubic polynomial; see Eq. (1.16).

After standard introduction of dimensionless variables, the equation under study takes the following form:

$$u_t = u_{xx} - \beta u + (1+\beta)u^2 - u^3 \, . \tag{3.39}$$

It can be shown that Eq. (3.39) has an exact solution in the following form:

$$u(x,t) = \frac{\beta \exp(\lambda_1\xi_1) + \exp(\lambda_2\xi_2)}{1 + \exp(\lambda_1\xi_1) + \exp(\lambda_2\xi_2)} \tag{3.40}$$

where $\xi_i = x - n_i t + \xi_{0i}$, $n_i = \sqrt{2}(1+\beta) - 3\lambda_i$, $i = 1, 2$,

$$\lambda_1 = \frac{\beta}{\sqrt{2}}, \quad \lambda_2 = \frac{1}{\sqrt{2}}, \tag{3.41}$$

and ξ_{01}, ξ_{02} are arbitrary constants.

There are different ways to arrive at (3.40). The simplest way to check that it gives a solution to Eq. (3.39) is, of course, through straightforward substitution. However, that gives no information regarding how the solution arises.

Kawahara and Tanaka (1983) obtained (3.40) by using a formal perturbation scheme. Later, it was shown by Danilov and Subochev (1991) that (3.40) can be obtained by expanding the solution into exponential series.

In this section, however, we consider another approach which is somewhat less laborious and can be applied to some other nonlinear models as well. In particular, in Chapters 4 and 6 we apply the same approach to study the interplay between diffusion and advection and the impact of inter-species interactions.

Let us introduce a new variable $w(x,t)$ defined by the following equation:

$$u(x,t) = \mu \, \frac{w_x}{w+\sigma} \tag{3.42}$$

where $\mu \neq 0$ is a coefficient (the case $\mu = 0$ would correspond to the trivial solution $u(x,t) \equiv 0$) and σ is a constant. In so far we are, for biological reasons, primarily interested in bounded solutions of Eq. (3.39), σ is included into the denominator of (3.42) in order to avoid singularities. If we assume that the function w is semi-bounded, i.e., there exists a certain $\bar{w}$ that either (i) $w(x,t) \leq \bar{w}$ or (ii) $w(x,t) \geq \bar{w}$ for $\forall\ x,t$, then constant σ can have an arbitrary value under the constraint $\sigma < -\bar{w}$ or $\sigma > -\bar{w}$ corresponding to the cases (i) and (ii), respectively.

Substitution of (3.42) to (3.39) leads to the following equation:

$$\begin{aligned} & \left[(2-\mu^2)w_x^3\right] (\ w+\sigma)^{-3} \\ & \quad + w_x \left[w_t - 3w_{xx} + (1+\beta)\mu w_x\right] (w+\sigma)^{-2} \\ & \quad + \left[w_{xxx} - \beta w_x - w_{xt}\right] (w+\sigma)^{-1} \ = \ 0 \end{aligned} \tag{3.43}$$

which, since constant σ is (nearly) arbitrary and different powers of $(w+\sigma)$ are linearly independent, is equivalent to the following system:

$$w_{xt} = w_{xxx} - \beta w_x \ , \tag{3.44}$$

$$w_t = 3w_{xx} - (1+\beta)\mu w_x \ , \tag{3.45}$$

$$\mu = \pm\sqrt{2} \ . \tag{3.46}$$

Without loss of generality, we choose plus in Eq. (3.46) (minus would correspond to the change $x \to -x$). Taking the partial derivative of Eq. (3.45) with respect to x in order to eliminate w_{xt} from Eq. (3.44), then from the system (3.44–3.46) we arrive at

$$w_{xxx} - \frac{1+\beta}{\sqrt{2}}\, w_{xx} + \frac{\beta}{2}\, w_x \ = \ 0 \ , \tag{3.47}$$

$$w_t \ = \ 3w_{xx} - \sqrt{2}(1+\beta)w_x \ . \tag{3.48}$$

Therefore, as before with the Burgers equation, the Cole–Hopf transformation linearizes the diffusion-reaction equation (3.39). However, the transformation now leads to a system of two linear partial differential equations, not to a single one. Consistency of the system implies certain constraints on the solution structure. As a result, in this case we are not able to obtain the general solution. Instead, we are going to obtain a *special solution* describing formation and propagation of a population front.

The solution of linear equation (3.47) has the following form:

$$w(x,t) = f_0(t) + f_1(t)e^{\lambda_1 x} + f_2(t)e^{\lambda_2 x} \tag{3.49}$$

where $\lambda_{1,2}$ are the roots of the square polynomial:

$$\lambda^2 - \frac{1+\beta}{\sqrt{2}}\,\lambda + \frac{\beta}{2} = 0 \tag{3.50}$$

and thus $\lambda_{1,2}$ are given by (3.41).

To obtain functions $f_{0,1,2}$, we substitute Eq. (3.49) into (3.48). That leads to the following result:

$$f_0(t) = C_0\ , \quad f_i(t) = C_i e^{\gamma_i t} \tag{3.51}$$

where $\gamma_i = 3\lambda_i^2 - \sqrt{2}(1+\beta)\lambda_i,\ i = 1, 2$, and $C_{0,1,2}$ are arbitrary constants. Note that the form of $w(x,t)$ defined by Eqs. (3.49–3.51) appears to be in agreement with our earlier assumption about semi-boundedness of w; thus our analysis has been consistent.

From (3.42), (3.49) and (3.51), we obtain:

$$u(x,t) = \frac{\sqrt{2}\ [C_1\lambda_1 \exp(\lambda_1 x + \gamma_1 t) + C_2\lambda_2 \exp(\lambda_2 x + \gamma_2 t)]}{(C_0+\sigma) + C_1 \exp(\lambda_1 x + \gamma_1 t) + C_2 \exp(\lambda_2 x + \gamma_2 t)}\ . \tag{3.52}$$

Obviously, for $u(x,t)$ to be positive, it is necessary that $C_0 + \sigma$, C_1 and C_2 have the same sign, i.e., either $C_0 + \sigma > 0,\ C_{1,2} > 0$ or $C_0 + \sigma < 0,\ C_{1,2} < 0$. Thus, introducing new constants as $\xi_{0i} = (1/\lambda_i)\ln(C_i/[C_0+\sigma]),\ i = 1, 2$ and taking into account that $\sqrt{2}\lambda_{1,2} = u_{1,2}$ (where $u_1 = \beta$ and $u_2 = 1$ are the steady states of the spatially homogeneous system), from Eq. (3.52) we finally arrive at (3.40):

$$u(x,t) = \frac{\beta \exp(\lambda_1\xi_1) + \exp(\lambda_2\xi_2)}{1 + \exp(\lambda_1\xi_1) + \exp(\lambda_2\xi_2)}\ .$$

Note that, since $u(x,t)$ must be nonnegative for any x and t, solution (3.40) is valid only for $\beta \geq 0$.

Clearly, solution (3.40) corresponds to the following conditions at infinity: $u(x \to -\infty, t) = 0,\ u(x \to \infty, t) = 1$. The spatial distribution of the population as given by (3.40) for different times is shown in Fig. 3.3. For t not large, it describes a "decay" of the steady unstable state $u = \beta$ through the interaction of two traveling waves propagating towards each other.

Since $\lambda_1 < \lambda_2$, it is readily seen that, in the large-time limit (when the transients die out, cf. curves 4 and 5 in Fig. 3.3) or for suitable values of

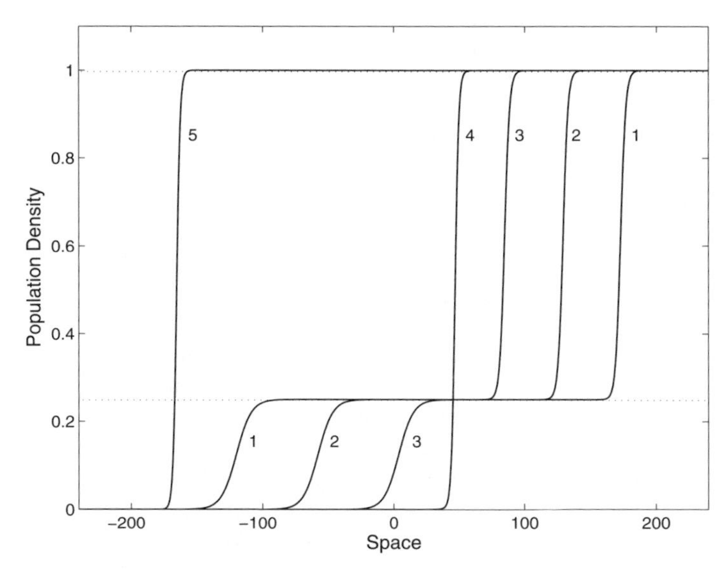

FIGURE 3.3: Evolution of the population density according to the exact solution (3.40) of Eq. (3.39): curves 1 to 4 for $t = 0$, $t = 50$, $t = 100$ and $t = 150$, respectively, curve 5 for $t = 750$. The dotted lines show the steady states. Parameters are: $\beta = 0.25$, $\phi_1 = 120$, $\phi_2 = -100$.

ϕ_1, ϕ_2, solution (3.40) reduces to

$$u(x,t) \simeq \frac{\exp(\lambda_2 \xi_2)}{1 + \exp(\lambda_2 \xi_2)} \tag{3.53}$$

describing a traveling population front propagating with the speed n_2 given by the following equation:

$$n_2 \;=\; \sqrt{2}(1+\beta) - \frac{3}{\sqrt{2}} \;=\; -\frac{1}{\sqrt{2}}(1-2\beta)\,. \tag{3.54}$$

Thus, the direction of the propagation can be either positive or negative:

$$n_2 \;<\; 0 \text{ for } \beta < \frac{1}{2} \quad \text{(the front propagates to the left)}, \tag{3.55}$$

$$n_2 \;>\; 0 \text{ for } \beta > \frac{1}{2} \quad \text{(the front propagates to the right)}. \tag{3.56}$$

Under condition (3.55) the front propagates to the region where the species is absent, which corresponds to species invasion, cf. Fig. 3.3; under condition (3.56) the front propagates to the region where the species is at its carrying capacity, which corresponds to species retreat. Conditions (3.55–3.56) are in full agreement with more general mathematical considerations, cf. (2.22–2.23).

In conclusion, we want to mention that, although solution (3.40) formally corresponds to a specific initial condition (which is immediately obtained from

(3.40) setting $t = 0$, cf. curve 1 in Fig. 3.3), it actually arises as a result of convergence of initial conditions from a wide class; for more details see Ognev et al. (1995).

3.3 Method of piecewise linear approximation

In the previous sections we saw that the Cole–Hopf transformation can linearize some nonlinear partial differential equation and thus makes them analytically solvable. Another approach that leads to exactly solvable models by means of linearization is based on a heuristic idea that any smooth curve can be approximated by a broken line. Concerning the single species model (2.14), the source of nonlinearity is the term $F(u)$ describing the population growth rate. Having approximated $F(u)$ with a broken line, e.g., see Fig. 3.4, instead of one nonlinear equation (2.14) we obtain a few linear ones, each of them describing the dynamics of given population in spatial domain(s) where the population density lays within the corresponding range. At the boundaries between the domains, the "local" solutions must match each other in order to provide a global continuous solution. Although this approach may seem somewhat naive from the mathematical point of view, in some cases it produces reasonable results helping to understand the properties of relevant traveling wave solutions.

Under the piecewise linear approximation, the number of the equations coincides with the number of the segments in the broken line. An attempt to use a many-segment line in order to provide a better approximation of the nonlinear function $F(u)$ normally results in a solution of very complicated form that can hardly be of any practical use. Also, the amount and tediousness of calculations increase dramatically. In contrast, as we are going to demonstrate in this section, a coarse approximation with only a two- or three-segment line can lead to solutions with biologically reasonable properties.

3.3.1 Exact solution for a population with logistic growth

The first known example of the piecewise linear approximation as applied to diffusion-reaction equations was given by Jones and Sleeman (1983). Let us consider the following equation of population dynamics:

$$u_t(x,t) = u_{xx} + F(u) \tag{3.57}$$

(in appropriately chosen dimensionless variables). We are going to look for a traveling wave solution of this equation assuming that the population exhibits

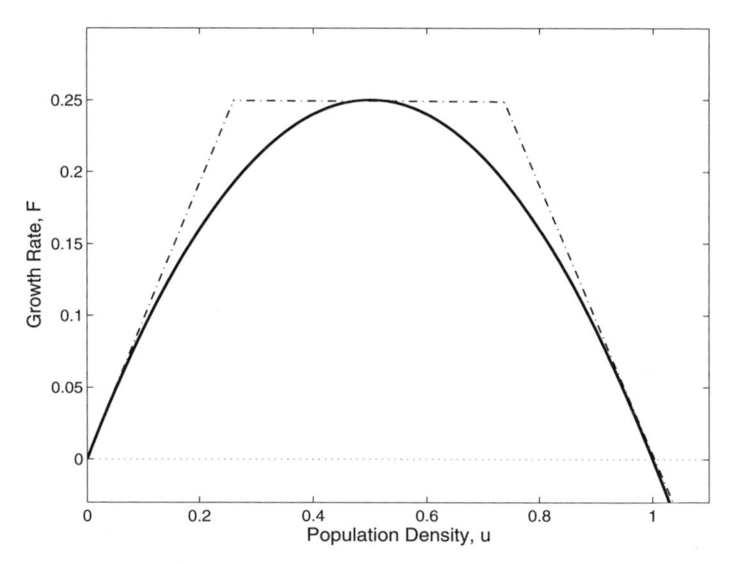

FIGURE 3.4: An example of piecewise linear approximation of the population growth rate.

logistic growth. Then, from (3.57) we obtain:

$$c\frac{du}{d\xi} + \frac{d^2u}{d\xi^2} + F(u) = 0 \quad \text{where} \quad F(u) = u(1-u) \tag{3.58}$$

and c is the speed of the wave.

We consider the following approximation for the growth function:

$$F(u) = u \quad \text{for} \quad 0 \leq u \leq \frac{1}{2} \quad \text{(region A)} \ , \tag{3.59}$$

$$F(u) = 1 - u \quad \text{for} \quad \frac{1}{2} \leq u \leq 1 \quad \text{(region B)} \ . \tag{3.60}$$

For convenience, we will refer to the region where u satisfies inequality (3.59) as region A and we will refer to the region where u satisfies (3.60) as region B.

The boundary conditions that correspond to traveling fronts should be the carrying capacity at the one end and zero at the other end. Assuming without a loss of generality that the invasive species spreads from left to right, we have $u(-\infty) = 1$ and $u(\infty) = 0$.

In region A, Eq. (3.58) takes the following form:

$$c\frac{du}{d\xi} + \frac{d^2u}{d\xi^2} + u = 0 \tag{3.61}$$

where the eigenvalues are given by the following equation:

$$\lambda_{1,2} = \frac{1}{2}\left(-c \pm \sqrt{c^2 - 4}\right) . \tag{3.62}$$

Correspondingly, in region B we have

$$c\frac{du}{d\xi} + \frac{d^2u}{d\xi^2} + 1 - u = 0 \tag{3.63}$$

with the following equation for the eigenvalues:

$$\mu_{1,2} = \frac{1}{2}\left(-c \pm \sqrt{c^2+4}\right) . \tag{3.64}$$

Further analysis depends on whether $c = 2$ or $c > 2$. In the case $c = 2$, in region A, both eigenvalues are negative and equal to -1. Correspondingly, the solution has the following form:

$$u_+(\xi) = A_1 \exp(-\xi) + B_1\xi \exp(-\xi) . \tag{3.65}$$

In region B, the solution is:

$$u_-(\xi) = 1 + A_2 \exp([-1+\sqrt{2}]\xi) + B_2 \exp([-1-\sqrt{2}]\xi) . \tag{3.66}$$

Here the four constants A_1, A_2, B_1 and B_2 are to be determined. Note that, due to the conditions at infinity chosen above, region B includes large negative values of ξ. Taking into account that the solution must be bounded, we immediately obtain that $B_2 = 0$.

To obtain remaining constants, we take into account that at the boundary between the regions the solutions $u_+(\xi)$ and $u_-(\xi)$ and their first derivative must match each other so that

$$u_+(\xi_0) = u_-(\xi_0) , \tag{3.67}$$

$$\frac{du_+(\xi_0)}{d\xi} = \frac{du_-(\xi_0)}{d\xi} \tag{3.68}$$

where ξ_0 is given by

$$u(\xi_0) = \frac{1}{2} , \tag{3.69}$$

cf. Eqs. (3.59–3.60). In a general case, there can be more than one point with the meaning of ξ_0. However, for a diffusion-reaction equation with logistic growth the monotonousness of the traveling fronts can be proved rigorously (Zeldovich and Barenblatt, 1959).

Let us also mention that, since the original equation does not include ξ explicitly, the solution must be invariant to translations. It means that the value of ξ_0 can be chosen arbitrary. For convenience, we let $\xi_0 = 0$. Then, from (3.67–3.69) we obtain that $A_1 = 1/2$, $A_2 = -1/2$ and $B_1 = (2-\sqrt{2})/2$, so that the solution is:

$$u(\xi) = \frac{1}{2}\exp(-\xi) + \left(1 - \frac{1}{\sqrt{2}}\right)\xi \exp(-\xi) \tag{3.70}$$

in region A ($\xi > 0$), and

$$u(\xi) = 1 - \frac{1}{2}\exp([\sqrt{2} - 1]\xi) \quad (3.71)$$

in region B ($\xi < 0$).

In the case $c > 2$, all eigenvalues are different and the solution has the following general form

$$u_+(\xi) = A_1 \exp(\lambda_1 \xi) + B_1 \exp(\lambda_2 \xi) \quad (3.72)$$

$$u_-(\xi) = 1 + A_2 \exp(\mu_1 \xi) + B_2 \exp(\mu_2 \xi) \quad (3.73)$$

in regions A and B, respectively. Applying the same procedure as above to find the constants A_1, A_2, B_1 and B_2, after somewhat more laborious calculations we arrive at the following solution:

$$u(\xi) = \frac{(\sqrt{c^2-4} + 2c - \sqrt{c^2+4})}{4\sqrt{c^2-4}} \exp\left(-\frac{1}{2}[c - \sqrt{c^2-4}]\xi\right)$$

$$+ \frac{(\sqrt{c^2-4} - 2c + \sqrt{c^2+4})}{4\sqrt{c^2-4}} \exp\left(-\frac{1}{2}[c + \sqrt{c^2+4}]\xi\right) \quad (3.74)$$

in region A ($\xi > 0$), and

$$u(\xi) = 1 - \frac{1}{2}\exp\left(-\frac{1}{2}[c - \sqrt{c^2+4}]\xi\right) \quad (3.75)$$

in region B ($\xi < 0$).

Figure 3.5 shows the traveling wave profiles obtained for $c = 2$ (solid curve 1) and for $c = 2.5$ (dashed-and-dotted curve 2). It is readily seen that lower speed corresponds to steeper slope. This is in full agreement with the results of more general considerations, cf. Murray (1989), Volpert et al. (1994). Let us note that the case $c = 2$ is biologically more realistic: the initial conditions of compact support typical for biological invasions always lead to traveling waves propagating with the minimum possible speed. However, the fact which is important for prospective applications is that the width of the transition region does not differ much between these two cases. In particular, it means that an exact solution obtained for a nonrealistic value of speed can be used to approximate the real wave profile. Another example of that kind will be considered in Section 3.4.1.

3.3.2 Exact solution for a population with a strong Allee effect

Spatiotemporal dynamics of populations with the Allee effect has been attracting a lot of attention recently. However, the origin of the Allee effect is

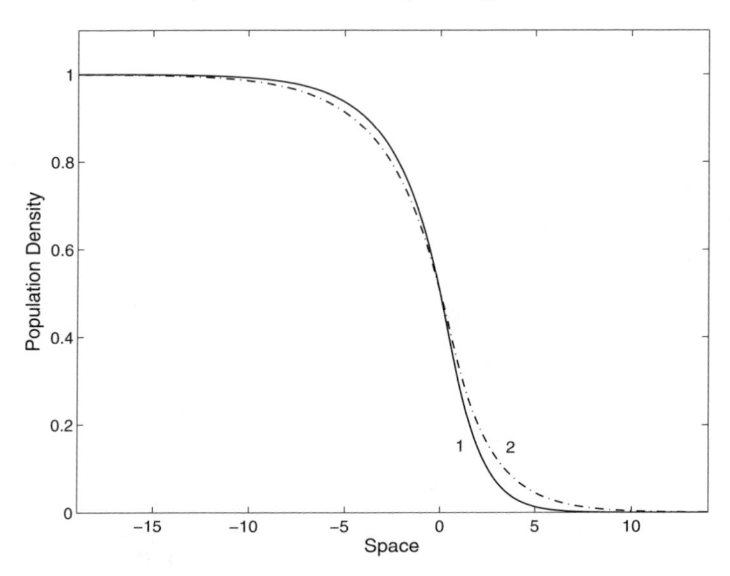

FIGURE 3.5: Exact traveling wave solution of Eq. (3.57) with the growth rate given by (3.59–3.60) obtained for $c = 2$ (solid curve 1) and $c = 2.5$ (dashed-and-dotted curve 2).

still understood rather poorly. Although a number of biological and environmental factors have been identified as its possible source, e.g., see Courchamp et al. (1999), the mechanistic theories are often lacking and that leaves many questions open. In particular, it is not clear whether the impact of the Allee effect exhibits itself for intermediate and large values of the population density where the effect of intraspecific competition is essential. The answer to this question affects the choice of the growth rate parameterization and thus, to a certain extent, the properties of the model. For instance, the usual parameterization of the growth rate with a cubic polynomial assumes implicitly that the impact of the Allee effect and that of intraspecific competition interfere for intermediate values of population density. In contrast, we can assume that this is not always true, and that there are two ranges: for small u only the Allee effect is important and the impact of intraspecific competition can be neglected, and for large u intraspecific competition is important but the Allee effect can be neglected. We then assume that the transition zone between the two ranges is narrow so that the properties of the growth rate experience a "jump" from one type of density-dependence to the other. That may justify the piecewise linear approximation (see Fig. 3.6), which makes some relevant models exactly solvable.

According to the above arguments, let us consider the following approximation of the growth function:

$$F(u) = -\alpha u + \delta\theta(u - u_A) \tag{3.76}$$

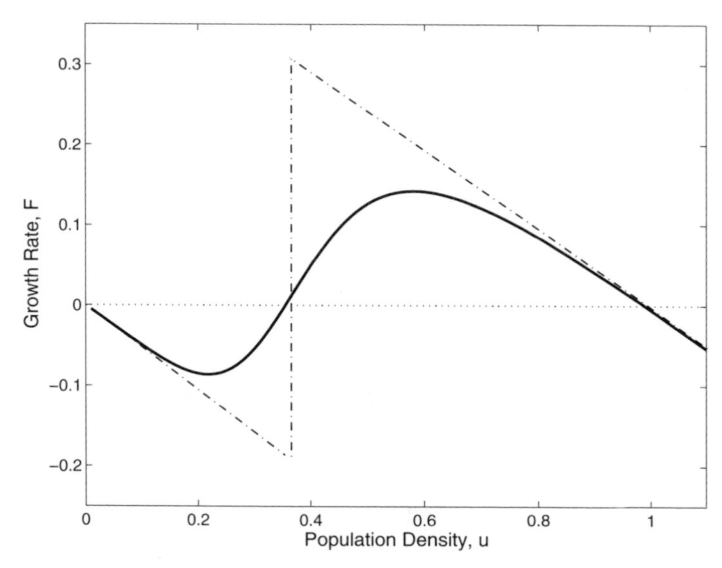

FIGURE 3.6: A sketch of piecewise linear approximation of the population growth rate in case of the strong Allee effect.

where α is the rate of population decay at small population density, u_A describes the position of the transition zone, δ is the magnitude of the "jump" and $\theta(u)$ is the Heaviside function, i.e.,

$$\theta(u) = 0 \text{ for } u < 0 \quad \text{and} \quad \theta(u) = 1 \text{ for } u \geq 0 \ . \tag{3.77}$$

Formally, the single-species model (2.14) with (3.76) depends on four parameters D, α, δ and u_A. However, introducing dimensionless variables

$$t' = t\alpha, \quad x' = x(\alpha/D)^{1/2}, \quad u' = u\alpha/\delta \tag{3.78}$$

(where δ/α is the population carrying capacity so that $F(\delta/\alpha) = 0$) and omitting primes for convenience, we arrive at the equation

$$u_t = u_{xx} - u + \theta(u - \beta) \tag{3.79}$$

which depends on a single parameter $\beta = u_A\alpha/\delta$.

Correspondingly, the traveling front of invasive species arises as a solution of the following equation:

$$c\frac{du}{d\xi} + \frac{d^2u}{d\xi^2} - u + \theta(u - \beta) = 0 \tag{3.80}$$

with the conditions at infinity as $u(-\infty) = 1$ and $u(\infty) = 0$.

Assuming that the wave profile is a monotonous function of ξ, the whole space is split into two domains, i.e., region A where $\xi > \xi_0$ and $u < \beta$, and region B where $\xi < \xi_0$ and $u > \beta$. The marginal point ξ_0 is defined by the relation $u(\xi_0) = \beta$.

In region A, from (3.80) we obtain:

$$\frac{d^2u}{d\xi^2} + c\frac{du}{d\xi} - u = 0 . \tag{3.81}$$

The eigenvalues are given as

$$\lambda_{\pm} = \frac{1}{2}\left(-c \pm \sqrt{c^2+4}\right) \tag{3.82}$$

where $\lambda_+ > 0 > \lambda_-$, and the solution has the following form:

$$u_+(\xi) = A_+ \exp(\lambda_+\xi) + B_+ \exp(\lambda_-\xi) . \tag{3.83}$$

Taking into account boundedness of the solution at infinity, we obtain $A_+ = 0$.

In region B, (3.80) takes the form

$$\frac{d^2u}{d\xi^2} + c\frac{du}{d\xi} + 1 - u = 0 . \tag{3.84}$$

It is straightforward to check that in this case the eigenvalues are given again by the same equation (3.82) so that the solution is:

$$u_-(\xi) = 1 + A_- \exp(\lambda_+\xi) + B_- \exp(\lambda_-\xi) \tag{3.85}$$

where $B_- = 0$ from the condition of solution boundedness at infinity.

Thus, we arrive at:

$$u_+(\xi) = B_+ \exp(\lambda_-\xi) \quad \text{for} \quad \xi > 0, \tag{3.86}$$

$$u_-(\xi) = 1 + A_- \exp(\lambda_+\xi) \quad \text{for} \quad \xi < 0. \tag{3.87}$$

To obtain remaining constants and the speed of the wave, we have the conditions of solution matching at the boundary between the regions:

$$u_+(\xi_0) = u_-(\xi_0) , \qquad \frac{du_+(\xi_0)}{d\xi} = \frac{du_-(\xi_0)}{d\xi} \tag{3.88}$$

where ξ_0 is given by $u(\xi_0) = \beta$.

From (3.88), we obtain:

$$B_+ \exp(\lambda_-\xi_0) = 1 + A_- \exp(\lambda_+\xi_0) \tag{3.89}$$

and

$$\lambda_- B_+ \exp(\lambda_-\xi_0) = \lambda_+ A_- \exp(\lambda_+\xi_0). \tag{3.90}$$

Accounting for the translation invariance of Eq. (3.80), ξ_0 can be arbitrary. Setting $\xi_0 = 0$, we immediately obtain $B_+ = \beta$ and $A_- = \beta - 1$. Thus, the solution to (3.80) is

$$u_+(\xi) = \beta \exp(\lambda_- \xi) \tag{3.91}$$

for $\xi > 0$, and

$$u_-(\xi) = 1 - (1 - \beta) \exp(\lambda_+ \xi) \tag{3.92}$$

for $\xi < 0$, where speed c is yet to be determined.

To obtain the speed, we have Eq. (3.90) which we have not used yet. With other constants now known, it reads as follows:

$$\lambda_- \beta = \lambda_+ (\beta - 1). \tag{3.93}$$

From (3.93), after a little algebra we obtain:

$$c = \frac{2z}{\sqrt{1 - z^2}} \tag{3.94}$$

where $z = 1 - 2\beta$.

The plot of speed dependence on β is shown in Fig. 3.7. Thus, $c = 0$ for $\beta = 0.5$ which is in full agreement with the results of a more general analysis, cf. (2.22). This coincidence looks encouraging and it is plausible that, in spite of the rather coarse approximation of the real growth function by means of (3.76), the solution (3.91–3.92) with (3.94) provides a good approximation to the traveling wave solution of the original equation in a certain parameter range.

Surprisingly, the speed goes to infinity for $\beta \to 0$ and for $\beta \to 1$. Taking into account the fact that the condition of wave blocking is given by the integral of $F(u)$ (see (2.22)), that seems rather counterintuitive because in both limiting cases the integral remains finite. It should be also mentioned that, in the case that growth function $F(u)$ is described by a cubic polynomial, the speed remains finite for all parameter values, cf. (2.20).

Let us note, however, that the appearance of infinite speed for $\beta \to 0$ becomes intuitively clear as soon as we recall that, from biological reasons, $F(u) = uf(u)$ where $f(u)$ is the per capita growth rate. Function $f(u)$ corresponding to $F(u)$ given by (3.76) is shown in Fig. 3.8. Thus, for $\beta \to 0$, the per capita growth rate tends to infinity for small u; naturally enough, it results in the unboundedly increasing wave speed. Note that similar behavior was also observed in another model where the per capita growth rate has singularity at $u = 0$; see Section 4.2.

This simple heuristic analysis seems to bring an important message to a more general theory of diffusion-reaction equations. The fact that, in a system with the Allee effect, the wave blocking condition is given by integral relation

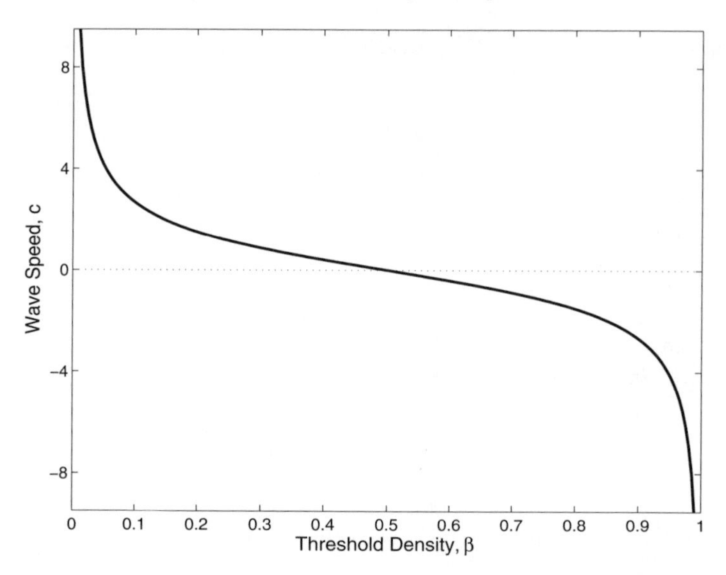

FIGURE 3.7: The speed of traveling population front for different values of threshold density β as given by the exact solution (3.91–3.92) of Eq. (3.79).

(2.22) inspired several authors to look for a general equation for the wave speed in the following form:

$$c = \phi(M) \quad \text{where} \quad M = \int_0^1 F(u)du. \tag{3.95}$$

The above results indicate that this hypothesis is unlikely to be justified. While for very small M this approach looks reasonable, cf. (2.23), in a wider parameter range it will hardly suffice. For larger deviations of M from 0, the details of function $F(u)$ are likely to become important.

3.4 Exact solutions of a generalized Fisher equation

In this section, we will consider two other methods to obtain exact solutions to diffusion-reaction equations using as an example the Fisher equation with a generalized growth function:

$$u_t(x,t) = u_{xx} + u(1 - u^q) \tag{3.96}$$

(in dimensionless variables) where $q > 0$ is a parameter. For $q = 1$, Eq. (3.96) coincides with the 1-D version of the "classical" Fisher equation given by (2.11).

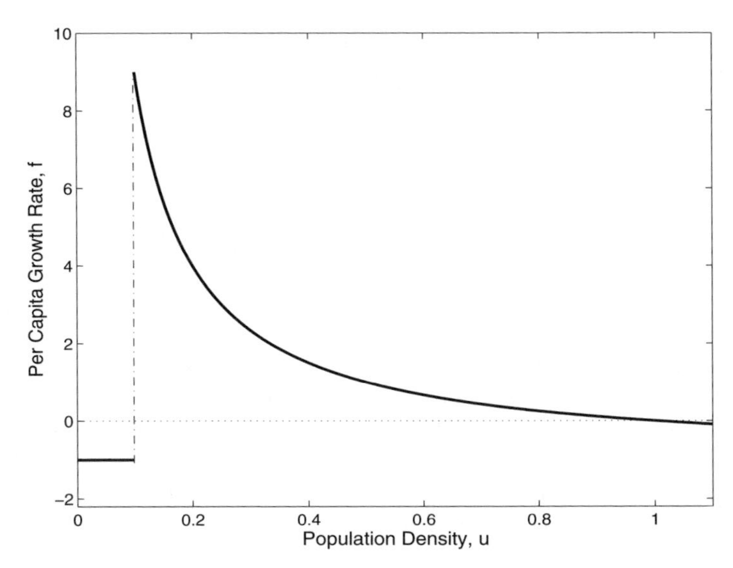

FIGURE 3.8: Density-dependence of the per capita population growth rate corresponding to piecewise linear approximation of $F(u)$ in case of the strong Allee effect.

3.4.1 Ansatz

In traveling wave variables, $u(x,t) = U(\xi)$ where $\xi = x - ct$, the generalized Fisher equation (3.96) reads as follows:

$$c\frac{dU}{d\xi} + \frac{d^2U}{d\xi^2} + U(1 - U^q) = 0 \ . \tag{3.97}$$

Having studied some asymptotic properties of Eq. (3.97) for $q = 1$, Murray (1989) noticed that its solution can be expanded into a power series with respect to $1/c$ where the main term is:

$$u(x,t) \quad = \quad U(\xi) \quad = \quad \frac{1}{1 + \exp(\xi/c)}. \tag{3.98}$$

That gives an idea that solutions in a more general case given by Eq. (3.97) with $q \neq 1$ may have a similar structure. Thus, we can introduce an *ansatz*, i.e., to look for an exact traveling wave solution of the generalized Fisher equation in a form inspired by (3.98).

To be more specific, we are going to look for a solution of (3.97) in the following form:

$$U(\xi) = \frac{1}{(1 + \exp(b\xi))^s} \ , \tag{3.99}$$

where positive constants b, s and the wave speed c are to be defined. Here "to be defined" implies that the ansatz is a suitable one; otherwise constants can not be obtained.

Having substituted (3.99) to (3.97), after standard transformations we arrive at

$$\begin{aligned} 1 + \left[s(s+1)b^2 - sb(b+c) + 1\right] \exp(2b\xi) \\ + \left[2 - sb(b+c)\right] \exp(b\xi) \\ - \left[1 + \exp(b\xi)\right]^{2-sq} &= 0 \,. \end{aligned} \tag{3.100}$$

The idea of obtaining the constants is as follows: since different powers of $\exp(b\xi)$ are linearly independent, Eq. (3.100) holds if and only if the coefficients at each power of $\exp(b\xi)$ are equal to zero. However, in order for Eq. (3.100) not to be overdetermined, the last term must be "congenial" to the others, i.e., it must not contain powers of $\exp(b\xi)$ others than 0, 1 and 2. That implies that $2 - sq = 2,\ 1$ or 0 so that

$$sq = 0 \,, \quad s = \frac{1}{q} \text{ or } \frac{2}{q} \,, \tag{3.101}$$

correspondingly.

Obviously, the first of these equations does not provide any suitable values since both s and q must be positive. Let us consider $sq = 1$. From (3.100), we then obtain:

$$\begin{aligned} &1 - sb(b+c) = 0, \\ &s(s+1)b^2 - sb(b+c) + 1 = 0 \end{aligned}$$

which gives $s(s+1)b^2 = 0$ and thus $b = 0$. This is of no use either since b must be positive, cf. (3.99).

For the remaining case $sq = 2$, from (3.100) we obtain:

$$\begin{aligned} &2 - sb(b+c) = 0, \\ &s(s+1)b^2 - sb(b+c) + 1 = 0 \end{aligned}$$

which yields

$$b = \frac{1}{\sqrt{s(s+1)}}, \qquad c = \frac{2}{sb} - b \tag{3.102}$$

and, finally,

$$U(\xi) = \left(1 + e^{b\xi}\right)^{-2/q} \quad \text{where } b = \frac{q}{\sqrt{2(2+q)}} \tag{3.103}$$

where the speed of the wave is given as

$$c = \frac{q+4}{\sqrt{2(2+q)}} \,. \tag{3.104}$$

Equations (3.103–3.104) give an exact analytical solution of the generalized Fisher equation (3.96).

Now, the question is whether solution (3.103–3.104) is fully relevant to biological invasion. Note that the generalized growth function in Eq. (3.96) is still of the type described by conditions (1.8–1.10) and thus the speed of the population wave generated by finite initial conditions must be $c = 2$ for any value of q. Instead, the speed c given by (3.104) is an increasing function of q so that its minimum value $c_0 = 2$ is reached for $q = 0$ and $c_q > 2$ for any $q > 0$. The waves with the speed larger than 2 can arise only in case of initial conditions with special properties (see Section 7.2) and, as such, they are not biologically realistic. Let us note, however, that for values q small enough the speed predicted by (3.104) is actually very close to 2. For instance, for $q = 1$, $q = 2$ and $q = 3$, we obtain, respectively:

$$c_1 = \frac{5}{\sqrt{6}} \approx 2.04, \quad c_2 = \frac{3}{\sqrt{2}} \approx 2.12, \quad c_3 = \frac{7}{\sqrt{10}} \approx 2.21.$$

Thus, it can be expected that, at least for $q = 1$, solution (3.103) approximates the "real" traveling front very well. Indeed, Fig. 3.9 shows the wave profiles given by (3.103) for $q = 1$ (curve 1), $q = 2$ (curve 2) and $q = 3$ (curve 3) as well as the wave profile obtained by means of numerical solution of the Fisher equation with finite initial conditions (dashed-and-dotted curve). While the profiles obtained for $q = 2$ and $q = 3$ are somewhat steeper than the front obtained numerically, the profile given by (3.103) for $q = 1$ is practically indistinguishable from the "real" one.

In conclusion, we want to mention that our observation regarding the steepness of the wave profiles increasing with the value of q does not contradict to the conclusion that faster fronts cannot be steeper than slower ones (Murray, 1989). The latter result concerns the case when the fronts with different speed arise in the same equation, i.e., for the same $F(U)$. The profiles shown in Fig. 3.9 correspond to different growth functions and thus the "speed-steepness relation" simply does not apply. Moreover, the greater steepness is not, actually, an artifact of the nonrealistic speed value but just a consequence of the growth function properties: the profiles obtained in numerical simulations with $q = 2$ and $q = 3$ (not shown here) appear to be very close to the corresponding analytical solution.

3.4.2 * The Ablowitz–Zeppetella method

The method used in the previous section to obtain the exact solution of the generalized Fisher equation was essentially based on the idea that the solution should have a certain structure; see (3.99). As well as a suitable change of variables, cf. Section 3.1, this approach appears very fruitful in analytical studies. A serious drawback, however, is that the underlying information about the solution structure is often missing; in this situation, the successful choice of ansatz is mostly based on the intuition and experience of the researcher. Since these valuable properties cannot always be guaranteed, there is a need of complementary or alternative approaches. In this section, we will show how an exact traveling wave solution for the Fisher equation can be obtained without making the hypothesis (3.99).

The simple observation that lies at the background of this analysis is that the Fisher equation written in the traveling wave coordinates,

$$\frac{d^2U}{d\xi^2} + c\frac{dU}{d\xi} + U(1-U) = 0 \tag{3.105}$$

where c is the wave speed, when linearized around $U = 0$, has a solution in the following form:

$$U(\xi) = Ae^{\lambda_1 \xi} + Be^{\lambda_2 \xi} \tag{3.106}$$

where

$$\lambda_{1,2} = \frac{1}{2}\left(-c \pm \sqrt{c^2 - 4}\right) \tag{3.107}$$

if $c > 2$, and $\lambda_{1,2} = -1$ if $c = 2$.

The form of (3.106) reminds us of a more general fact that, in a certain domain around a given steady state in the phase plane of a given nonlinear

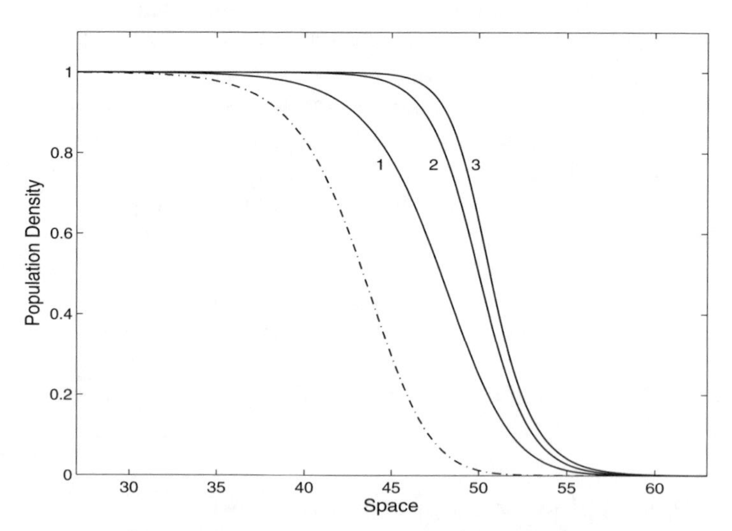

FIGURE 3.9: Exact traveling wave solution (3.103–3.104) of the generalized Fisher equation (3.96) obtained for different type of density-dependence and thus corresponding to different value of speed: curve 1 for $q = 1$ ($c = 2.04$), curve 2 for $q = 2$ ($c = 2.12$) and curve 3 for $q = 3$ ($c = 2.21$). The dashed-and-dotted curve shows numerical solution of Eq. (3.96) with $q = 1$ propagating with the minimum speed $c = 2$.

ordinary differential equation of the second order, its solution can be expanded into exponential series:

$$U(\xi) = \sum_{m+n\geq 1} A_{nm} \exp[(m\lambda_1 + n\lambda_2)\xi] \tag{3.108}$$

(cf. Lefschetz, 1963) where summation is done over all nonnegative whole numbers m and n. Apparently, the right-hand side of Eq. (3.106) corresponds to the first two terms of the sum (3.108).

Note that Eq. (3.105) has another steady state corresponding to $U = 1$ and thus an expansion similar to (3.108) can be written as well in a domain around $(1, 0)$. Assuming that the two domains overlap, the most general approach then would be to match the two series. However, due to the generality of the solution in the form of (3.108), this approach would unlikely lead to a solution in a closed form.

Alternatively, Ablowitz and Zeppetella (1979) tried to link Eq. (3.105) to a class of *integrable equations*, i.e., nonlinear equations that can be solved explicitly. The main idea of their analysis is that some of the equation's properties can be different for different values of c. The property that is of particular interest in this context is the so-called Painlevé property: an equation is said to be of Painlevé type in case it possesses, if considered as a function of a complex variable, only poles as "movable" (i.e., dependent on initial conditions) singularities. There is considerable evidence that a nonlinear ordinary differential equation is explicitly solvable if and only if it possesses the Painlevé property; see Ablowitz et al. (1978), Weiss et al. (1983) and Newell et al. (1987) for details and further reference.

Having done necessary calculations, Ablowitz and Zeppetella (1979) found that, indeed, Eq. (3.105) is of Painlevé type for $c = 5/\sqrt{6}$ (assuming c to be nonnegative). Using this value, from (3.107) we obtain that $\lambda_1 = -3/\sqrt{6}$ and $\lambda_2 = -2/\sqrt{6}$ so that the expansion (3.108) takes the following form:

$$U(\xi) = \sum_{n=2}^{\infty} A_n \exp\left(-\frac{n\xi}{\sqrt{6}}\right) . \tag{3.109}$$

Having substituted (3.109) into Eq. (3.105), after equating the coefficients of different powers of $\exp(-\xi/\sqrt{6})$ we obtain that

$$A_n = \frac{6}{(n-2)(n-3)} \sum_{j=2}^{n-2} A_j A_{n-j} , \quad n \geq 4 , \tag{3.110}$$

while the coefficients A_2 and A_3 remain arbitrary. Different choice of A_2 and A_3 generates different trajectories coming out of the steady state $(0, 0)$ in the phase plane of Eq. (3.105). Let us recall now that we are interested in the solution with $U(-\infty) = 1$, $U(\infty) = 0$ so that the relevant values must correspond to the trajectory arriving at $(1, 0)$.

Also, we expect that a successful choice of the coefficients should make it possible to write the solution in a closed form. Letting $A_2 = 1$ and $A_3 = -2$, from (3.110) we arrive at

$$A_n = (-1)^n (n-1), \quad n \geq 2 . \tag{3.111}$$

Denoting, for convenience, $\exp(-\xi/\sqrt{6}) = y$, from (3.109) and (3.111) we then obtain

$$U = \sum_{n=2}^{\infty} (-1)^n (n-1) y^n = y^2 \sum_{n=0}^{\infty} (-1)^n (n+1) y^n . \tag{3.112}$$

It is readily seen that the sum in the right-hand side of (3.112) gives the power-series expansion of $(1+y)^{-2}$; thus, (3.112) takes the following form:

$$U = \frac{y^2}{(1+y)^2} . \tag{3.113}$$

Finally, coming back to original variables, we obtain the exact solution:

$$U(\xi) = \left(1 + e^{z/\sqrt{6}}\right)^{-2} . \tag{3.114}$$

Apparently, solution (3.114) coincides with a more general solution (3.103) in the particular case $q = 1$. Let us emphasize, however, that, contrary to the previous method, in this section we did not use any a priori information about the solution structure.

3.5 More about ansatz

We have already seen that probably the easiest and the fastest way to obtain an exact solution is through using a relevant ansatz. In this section, we are going to develop the ideas introduced in Section 3.4.1 and give two more examples showing how ansatz is constructed and used in order to obtain biologically relevant exact solutions of diffusion-reaction equations.

Let us consider a generalized Fisher equation in traveling wave coordinates,

$$\frac{d^2U}{d\xi^2} + c\frac{dU}{d\xi} + F(U) = 0 \tag{3.115}$$

(in dimensionless units), c being the wave speed. We choose the following conditions at infinity: $U(-\infty) = 0$, $U(\infty) = 1$. Thus, species invasion corresponds to the front propagating to the left, i.e., against axis x.

By means of introducing a new variable, $\psi(U) = dU/d\xi$ (see Section 9.1), Eq. (3.115) takes the form

$$\psi(U)\frac{d\psi}{dU} + c\psi(U) + F(U) = 0 \tag{3.116}$$

where $\psi(U)$ must satisfy the following conditions:

$$\psi(0) \;=\; \psi(1) \;=\; 0. \tag{3.117}$$

The solution of Eq. (3.116) depends on the properties of function F. Thus, an appropriate form of ansatz can be sought "congenial" to the form of $F(U)$. In particular, let us consider the case when function F is a polynomial:

$$F(U) = U(a_0 + a_1 U + \ldots + a_{k-1}U^{k-1}) \tag{3.118}$$

where $a_0, \ldots, a_k$ are coefficients. That inspires an idea that ψ can be a polynomial as well.

The first step is to find out what the power of the polynomial can be. In order to do that, we substitute $\psi = U^m$ to (3.116–3.118) and try to match the leading powers of U:

$$U^m U^{m-1} - U^m \sim U^k \ . \tag{3.119}$$

Since $2m - 1 > m$ for any $m > 1$, we can neglect the second term in the left-hand side of (3.119). From (3.119) we then obtain that $2m - 1 = k$. Thus, the first conclusion we can make is that Eq. (3.116) with F(U) as (3.118) can only have polynomial solutions if k is an odd number.

For biological reasons, it is rather unlikely that the growth rate depends on the population density in a very complicated manner. Also, from the modeling standpoint, intricate density-dependence can hardly ever be justified because ecological data are usually of poor accuracy. Recalling the polynomial (3.118), it means that either k should not be large or most of the coefficients $a_0, \ldots, a_k$ are equal to zero.

Case A. Let us first consider the case $k = 3$; correspondingly,

$$F(U) = U(U - \beta)(1 - U) = \; -\beta U + (\beta + 1)U^2 - U^3, \tag{3.120}$$

cf. (1.16), where parameter β is not necessarily positive now. Taking into account (3.117), the relevant form of ansatz is

$$\psi = a(U - U^2) \tag{3.121}$$

where a is a coefficient to be determined. Substituting (3.121) to (3.116) and matching different powers of U, we obtain the following system:

$$\begin{aligned} a^2 + ca - \beta &= 0, \\ -3a^2 - ca + \beta + 1 &= 0, \\ 2a^2 - 1 &= 0. \end{aligned}$$

Here the last equation gives $a = \pm 1/\sqrt{2}$ where we choose plus since in our case $\psi = dU/d\xi > 0$ along the wave profile. The first and the second equations appear to be identical and provide the value of the wave speed:

$$c = -\frac{1}{\sqrt{2}}(1 - 2\beta) \tag{3.122}$$

which coincides with the result obtained earlier, cf. (2.20). The corresponding exact solution arises as a solution of the equation

$$\frac{dU}{d\xi} = \frac{1}{\sqrt{2}}(U - U^2) \tag{3.123}$$

from which we readily obtain

$$U(\xi) = \frac{\exp(\xi/\sqrt{2})}{1 + \exp(\xi/\sqrt{2})} \; . \tag{3.124}$$

Eq. (3.124) describes propagation of a traveling population front in the single-species model where the population growth is described by a cubic polynomial. Clearly, it gives the asymptotical solution to the "full" partial differential equation (3.39) in the large-time limit, cf. (3.53) and (3.124). Let us note, however, that (3.124) was obtained without any restriction on the value of β. As a matter of fact, it does not contain β at all and the solution depends on β only via the wave speed; see (3.122). As such, the solution (3.124) is applicable to the dynamics of populations with the weak Allee effect, $-1 < \beta \leq 0$, and to populations without Allee effect, $\beta \leq -1$.

In particular, for $\beta = -1$, (3.124) describes a traveling population front propagating with the speed $c = 3/\sqrt{2}$. The corresponding growth function is $F(U) = U(1 - U^2)$; thus, in this case (3.124) coincides with the solution of the generalized Fisher equation obtained for $q = 2$, cf. (3.96) and (3.103–3.104) (up to the change $x \to -x$ which correspond to the opposite choice of conditions at infinity).

Case B. Let us now consider a more general case when the maximum power of density-dependence in $F(U)$ is $2m - 1$ where $m > 2$. Following our strong belief that an adequate ecological model should not be too complicated (see also the comments above Eq. (3.120)), we assume that $F(U)$ does not actually include all powers of U between 1 and $2m - 1$. More specifically, we consider the following growth function:

$$F(U) = BU + AU^m + \kappa U^{2m-1} \tag{3.125}$$

where A, B and κ are coefficients.

The maximum power in the polynomial ansatz $\psi(U)$ is m; see (3.119). Let us consider

$$\psi = a(U - U^m) \; . \tag{3.126}$$

Although (3.126) is of a special form, it can be shown that the choice of ansatz in the form of a general polynomial of power m does not provide any new option compared to (3.126).

Substituting (3.126) into (3.116) with (3.125) and matching different powers of U, we arrive at the following equations:

$$a^2 + ca + B = 0, \tag{3.127}$$

$$-(m+1)a^2 - ca + A = 0, \tag{3.128}$$

$$ma^2 + \kappa = 0. \tag{3.129}$$

From the last equation, we get $a^2 = -\kappa/m$; this means that our approach is applicable only if $\kappa < 0$. Assuming that the population density U is scaled to be between 0 and 1, we let $\kappa = -1$. Then, from Eq. (3.129) we obtain:

$$a = \pm \frac{1}{\sqrt{m}} \tag{3.130}$$

where we choose plus because $dU/d\xi$ must be positive due to our choice of the conditions at infinity. The other two equations are now reduced to

$$c = -\sqrt{m}\left(B + \frac{1}{m}\right), \tag{3.131}$$

$$c = -\sqrt{m}\left((1-A) + \frac{1}{m}\right). \tag{3.132}$$

These equations are consistent only if $A + B = 1$. Choosing B as an independent parameter, we obtain that the speed of the wave is

$$c = -\frac{1}{\sqrt{m}}(1 + Bm) . \tag{3.133}$$

Respectively, the growth function (3.125) takes the form:

$$F(U) = BU + (1-B)U^m - U^{2m-1} . \tag{3.134}$$

Since the properties of $F(U)$ depend on parameter B, Eq. (3.115) with (3.134) describes a variety of ecological situations. It is not difficult to see that, for $B \geq 1$, function $F(U)$ is convex and thus corresponds to a population without Allee effect. In this case, the wave speed is always negative which indicates successful species invasion. For $0 \leq B < 1$, $F(U)$ is not convex any more but F remains positive for $0 < U < 1$ which corresponds to the weak Allee effect. Although the rate of spread is getting slower, the speed remains negative which means species invasion: the impact of weak Allee effect can slow it down but cannot block it. The range $B < 0$ corresponds to the strong Allee effect when the growth rate becomes negative for $0 < U < B$. In this case, the speed changes its sign at $B = -1/m$ so that invasion turns to retreat for $B < -1/m$.

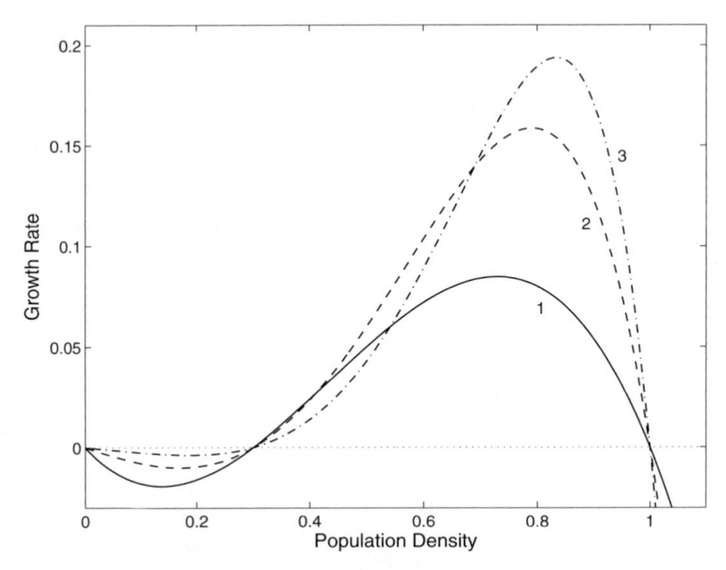

FIGURE 3.10: Population growth rate in case of a higher-order density-dependence, cf. (3.135): curves 1, 2 and 3 show $F(U)$ for $m = 2$, $m = 3$ and $m = 4$, respectively.

Apparently, function (3.134) provides a generalization of the usual parameterization of the Allee effect via a cubic polynomial to the case of a higher-order density-dependence. The resemblance between (3.120) and (3.134) becomes even more obvious if we focus on the populations with the strong Allee effect. In this case $B < 0$; then, introducing a new parameter β so that $B = -\beta^{m-1}$, the growth function (3.134) is written as

$$F(U) = U\left(U^{m-1} - \beta^{m-1}\right)\left(1 - U^{m-1}\right). \tag{3.135}$$

The one-parameter family of functions (3.135) is shown in Fig. 3.10 where curves 1, 2 and 3 correspond to $m = 2$, $m = 3$ and $m = 4$, respectively.

The exact solution describing the wave profile arises from the equation

$$\frac{dU}{d\xi} = \frac{1}{\sqrt{m}}(U - U^m)\,, \tag{3.136}$$

cf. (3.126), which is equivalent to

$$\int \frac{dU}{U - U^m} = \frac{\xi - \xi_0}{\sqrt{m}} \tag{3.137}$$

where ξ_0 is the integration constant. In order to calculate the integral in the left-hand side of (3.136), let us introduce a new variable y so that

$$y = U^{m-1}, \qquad U = y^{1/(m-1)}.$$

Correspondingly,

$$\int \frac{dU}{U - U^m} = \frac{1}{m-1} \int \frac{dy}{y(1-y)} = \frac{1}{m-1} \log \left| \frac{y}{1-y} \right| .$$

From (3.137), we then obtain

$$\log \left| \frac{y}{1-y} \right| = \frac{m-1}{\sqrt{m}} (\xi - \xi_0)$$

and, finally,

$$U(\xi) = \exp\left(\frac{\xi}{\sqrt{m}}\right) \left[1 + \exp\left(\frac{m-1}{\sqrt{m}}\xi\right)\right]^{-1/(m-1)} , \qquad (3.138)$$

where, for convenience, we have included ξ_0 into the definition of ξ. The exact traveling wave solution (3.138) where the wave speed is given by (3.133) provides an extension of the solution (3.124) with (3.122) to the case of higher-order density-dependence, cf. (3.120) and (3.134–3.135).

Using the approach described in this section, it appears possible to find exact traveling wave solutions also for a few somewhat more complicated cases, e.g., when the growth function is a quintic polynomial with all powers of U being present. However, the form of the exact solution appears rather cumbersome and, as such, not very instructive. Also, the biological implications of this case are not clear; see the comments above Eq. (3.120). Thus, we will not discuss it here; a relevant analysis can be found in Otwinowski et al. (1988) and Benguria and Depassier (1994).

Chapter 4

Single-species models

In this chapter, we consider a few more exactly solvable models of population dynamics relevant to biological invasion. Our purpose is to take into account some features of biological invasion that have not been addressed before and to further demonstrate that exactly solvable models make a valuable contribution to modeling and understanding species invasion. Trying to keep the models "as simple as possible (but not simpler)," here we focus on single-species models described by a single partial differential equation. To obtain the exact solutions, we mostly use the methods already described in the previous chapter, with some modifications where necessary.

When dealing with invasive species, a highly practical and theoretically important problem is the identification of factors that can either decrease or increase the rate of the species geographical spread. Although a number of such factors has been revealed and studied, there are many issues that are still poorly understood. In particular, the role of small-scale migrations is not clear. Such migrations can appear as a result of certain density-dependent behavioral responses, cf. "stratified diffusion" (Hengeveld, 1989). One obvious way to account for the density-dependence of the species motility is through considering density-dependent diffusion, i.e., changing the constant coefficient D to a function $D(u)$ in the corresponding diffusion-reaction equations. A few models of that kind will be considered in Chapter 5. Alternatively, in Section 4.1 we consider an approach that treats variable species motility by means of linking them to small-scale migrations. We consider the interplay between advection, migration and diffusive spread hampered by the strong Allee effect and show that the outcome of this interplay can be counterintuitive.

Geographical spread of invasive species typically takes place at a constant rate, i.e., the corresponding population front propagates with a constant speed. However, there also exists another pattern of spread when the front propagates with an increasing speed. These patterns are sometimes referred to as traveling and dispersive fronts, respectively, e.g., see Frantzen and van den Bosch (2000). The origin of the dispersive fronts is often seen in the fat-tailed dispersive kernels arising in the case of non-Gaussian diffusion; see Kot et al. (1996) and also Section 2.2 of this book. The mechanistic models explaining the nature of such kernels are still lacking, though, and that leaves many questions open. In this situation, the search for alternative mechanisms that can result in accelerating or dispersive waves becomes a challenging issue.

In Section 4.2 we show that accelerating fronts may correspond to a transient stage of species spread preceding the stage of constant-speed propagation. Remarkably, in a finite-size domain (which is always the case in real ecosystems) and for a sufficiently high population growth rate, the constant-speed stage may never be achieved.

Invasion of exotic species starts with its introduction and local establishment. Successful species establishment typically results in its geographical spread (but see Petrovskii et al., 2005b) while unsuccessful one leads to species extinction. Apart from many ecological and environmental factors that affect the outcome of the establishment stage (see Sakai et al. (2001) for details), there are purely dynamical aspects arising from the interplay between diffusion, population growth and mortality, especially mortality at low population densities. It seems intuitively clear that a large initial population size will likely lead to species invasion while a small population size will more probably result in extinction. In Section 4.3 we give a mathematically rigorous consideration of this problem and show that these two outcomes can be distinguished in terms of critical population density and/or critical radius of the infested domain.

4.1 Impact of advection and migration

Standard models of population dynamics described by diffusion-reaction equations are based on the assumption that the individual motion is random and isotropic in space. There is, however, another type of dynamics when the individuals exhibit a correlated motion towards a certain direction. The origin of this motion can be different, and there are at least two apparently different mechanisms resulting in the species transport. We will call "advection" the correlated motion caused by purely environmental factors such as wind in case of air-borne species or water current in case of water-borne species, and we will call "migration" the transport caused by biological interactions. Evidently, species transport due to the advection is density independent. As for migration, there are certain indications that its "intensity," i.e., either the speed of individual motion or the number of migrating individuals, can increase with the population density. To avoid ambiguousness, it should be mentioned that here we are not interested in the periodical return migrations which are typical for many bird and fish species. A point of interest is the migration occurring on a much smaller spatial scale when individuals or groups of individuals of the given species move from the regions with high population density towards the regions where the given species is either absent or exists at low population density. Although these small-scale migrations are often interpreted in terms of random motion, cf. "stratified diffusion," the phenomenon exhibits apparent spatial anisotropy. Interestingly, this mechanism of species dispersion usually comes to operation when the population density becomes sufficiently high.

Note that both random and correlated motion can result in blocking species invasion or even in turning it into retreat. For the correlated motion, this possibility is obvious: whether the species is actually invading or retreating depends on the direction of advective transport. For the random motion, i.e., diffusion, species retreat results from increasing Allee effect; see Sections 2.1 and 3.2.

In theoretical studies, the two types of motion, i.e., random or correlated, are usually considered separately. In a real ecological community, however, the individuals are likely involved in a combination of these two types of motion. An issue of interest is the interplay between these two types of motion from the standpoint of species invasion. In particular, can advection/migration block the species spreading in the case when the invasion would otherwise take place due to random motion?

Our analysis here is essentially based on the original paper by Petrovskii and Li (2003). We assume that the given species is involved both in random and correlated motion and consider the following 1-D model allowing for advection/migration and diffusion:

$$\frac{\partial u(x,t)}{\partial t} + \frac{\partial (Au)}{\partial x} = D\frac{\partial^2 u}{\partial x^2} + F(u) \ , \tag{4.1}$$

cf. (2.7), where A is positive in the case that advection/migration is going in the direction of axis x and negative otherwise.

We assume that the growth rate is damped by the Allee effect and choose the cubic polynomial parameterization for $F(u)$:

$$F(u) = \omega u(u - u_A)(K - u). \tag{4.2}$$

For the speed of migration, we assume that it is given by

$$A = A_0 + A_1 u \tag{4.3}$$

where A_0 and A_1 are parameters, A_0 is the speed of advection caused by environmental factors, e.g., by wind or water current, and $A_1 u$ is the speed of migration due to biological mechanisms. For convenience, we will call parameter A_1 the per capita migration speed.

Introducing dimensionless variables,

$$\tilde{u} = \frac{u}{K} \ , \quad \tilde{t} = t\omega K^2 \ , \quad \tilde{x} = x\sqrt{\frac{\omega K^2}{D}} \ , \tag{4.4}$$

and omitting tildes further on for notation simplicity, from (4.1–4.3), we obtain:

$$u_t + (a_0 + a_1 u)u_x = u_{xx} - \beta u + (1+\beta)u^2 - u^3 \tag{4.5}$$

where $\beta = u_A/K$, $a_0 = A_0 K^{-1}(\omega D)^{-1/2}$ and $a_1 = 2A_1(\omega D)^{-1/2}$ are dimensionless parameters, and subscripts x and t stand for the partial derivatives with respect to dimensionless space and time, respectively.

Eq. (4.5) is considered in the unbounded domain with the following conditions at infinity:

$$u(x \to -\infty, t) \;=\; 0, \qquad u(x \to +\infty, t) \;=\; 1, \tag{4.6}$$

i.e., invasion of alien species is going from right to left. Therefore, negative a_0, a_1 correspond to advection/migration enhancing species invasion and positive a_0, a_1 correspond to advection/migration hampering species invasion or enhancing species retreat.

An exact solution of Eq. (4.5) for $a_0 = a_1 = 0$ was obtained in Section 3.2 (see (3.40)) by means of using the generalized Cole-Hopf transformation. Now we are going to apply the same method in order to obtain an exact solution of the full advection-diffusion-reaction equation (4.5). We begin by considering the cases of advection and migration separately, proceeding then to a general case.

4.1.1 Advection

In the case that the speed of the species transport does not depend on the population density, e.g., when drifting with the wind, the dynamics of the population is described by the following equation:

$$u_t + a_0 u_x = u_{xx} - \beta u + (1+\beta)u^2 - u^3 \; . \tag{4.7}$$

Considering traveling wave coordinates, $(x, t) \to (\xi, t)$ where $\xi = x - a_0 t$, so that $u = \tilde{u}(\xi, t)$, Eq. (4.7) turns to

$$\tilde{u}_t = \tilde{u}_{\xi\xi} - \beta\tilde{u} + (1+\beta)\tilde{u}^2 - \tilde{u}^3 \; . \tag{4.8}$$

Eq. (4.8) coincides with (3.39) and thus the exact solution (3.40) gives also an exact solution of (4.8) with the obvious change $x \to \xi$. In particular, in the large-time limit, when the solution describes a single traveling population front, it reads as follows:

$$u(x,t) \;=\; \tilde{u}(x - a_0 t, t) \;\simeq\; \frac{\exp\{\lambda_2\left[x - (n_2 + a_0)t + \phi_2\right]\}}{1 + \exp\{\lambda_2\left[x - (n_2 + a_0)t + \phi_2\right]\}} \tag{4.9}$$

where $\lambda_2 = 1/\sqrt{2}$ and $n_2 = (1/\sqrt{2})(2\beta - 1)$. Clearly, $n_2 + a_0$ is the speed of the front so that $n_2 + a_0 < 0$ corresponds to the species invasion and $n_2 + a_0 > 0$ corresponds to the species retreat. Advection enhances the species invasion if $a_0 < 0$ and enhance the species retreat if $a_0 > 0$. Thus, the species will invade in spite of the counteracting impact of advection (e.g., cross-wind or cross-current) in case $n_2 < -a_0$, i.e., for

$$\frac{2\beta - 1}{\sqrt{2}} < -a_0 \tag{4.10}$$

which is equivalent to

$$\beta < \frac{1}{2}\left(1 - \sqrt{2}\, a_0\right). \tag{4.11}$$

Relation (4.11) has an intuitively clear meaning: the weaker the Allee effect is for a given population, the higher is its capability for invasion. In case we are restricted to the case of the strong Allee effect, β is positive while a_0 can be arbitrary. It means that a species affected by the strong Allee effect cannot invade in case the cross-wind is sufficiently strong, $a_0 > 1/\sqrt{2}$.

4.1.2 Density-dependent migration

Now we are going to consider the case when the density-independent advection caused by environmental factors is absent and migration takes place due to biological mechanisms which are assumed to be density-dependent. Then $a_0 = 0$ and from (4.5) we arrive at the following equation:

$$u_t + a_1 u u_x = u_{xx} - \beta u + (1+\beta)u^2 - u^3 \,. \tag{4.12}$$

Eq. (4.12) differs from (3.39) and the exact solution (3.40) is not immediately applicable. However, to obtain an exact solution of (4.12), we can try to make use of the approach described in Section 3.2.

Let us consider a new variable $p(x,t)$,

$$u(x,t) = \nu\, \frac{p_x}{p+\sigma}\,, \tag{4.13}$$

where ν is a coefficient and σ is a (nearly) arbitrary constant included in order to avoid singularities, cf. Section 3.2 for details. Having substituted Eq. (4.13) into (4.12), we arrive at the following system:

$$p_{xt} = p_{xxx} - \beta p_x \,, \tag{4.14}$$

$$p_t = (3 + a_1\nu)p_{xx} - (1+\beta)\nu p_x \,, \tag{4.15}$$

$$\nu = \frac{1}{2}\left(a_1 \pm \sqrt{a_1^2 + 8}\right). \tag{4.16}$$

Choosing plus in Eq. (4.16) without any loss of generality (minus would correspond to the change $a_1 \to -a_1,\ x \to -x$) and excluding p_{xt} from Eq. (4.14), from (4.14–4.16) we obtain:

$$(2 + a_1\nu)p_{xxx} - (1+\beta)\nu\, p_{xx} + \beta p_x \;=\; 0\,, \tag{4.17}$$

$$p_t \;=\; (3 + a_1\nu)p_{xx} - (1+\beta)\nu p_x \,. \tag{4.18}$$

The solution of the linear equation (4.17) has the following form:

$$p(x,t) = g_0(t) + g_1(t)e^{\omega_1 x} + g_2(t)e^{\omega_2 x} \tag{4.19}$$

where

$$\omega_1 = \frac{\beta}{\nu} , \quad \omega_2 = \frac{1}{\nu} . \tag{4.20}$$

Having substituted (4.19) into (4.18), we obtain:

$$g_0(t) = B_0 , \quad g_i(t) = B_i e^{\delta_i t} \tag{4.21}$$

where $\delta_i = (3+a_1\nu)\omega_i^2 - (1+\beta)\nu\omega_i$, $i = 1, 2$ and $B_{0,1,2}$ are arbitrary constants.

Considering Eq. (4.13) together with (4.19) and (4.21) and taking into account that coefficients $B_0 + \sigma$, B_1 and B_2 must have the same sign in order to provide positiveness of the solution, we arrive at the following exact solution of Eq. (4.12):

$$u(x,t) = \frac{\beta \exp(\omega_1\psi_1) + \exp(\omega_2\psi_2)}{1 + \exp(\omega_1\psi_1) + \exp(\omega_2\psi_2)} \tag{4.22}$$

where $\psi_i = x - q_i t + \psi_{i0}$, $q_i = (1+\beta)\nu - (3 + a_1\nu)\omega_i$, $i = 1, 2$ and ψ_{01}, ψ_{02} are arbitrary constants.

The structure of solution (4.22) is apparently similar to that of (3.40) and it has similar properties. In particular, since $\omega_1 < \omega_2$, in the large-time limit solution (4.22) describes a single traveling population front

$$u(x,t) \simeq \frac{\exp(\omega_2\psi_2)}{1 + \exp(\omega_2\psi_2)} \tag{4.23}$$

propagating with the speed

$$q_2 = (1+\beta)\nu - \frac{3 + a_1\nu}{\nu} . \tag{4.24}$$

From (4.24), after some standard transformations, we obtain

$$q_2 = \beta\nu - \frac{1}{\nu} \tag{4.25}$$

where $\nu = 0.5(a_1 + \sqrt{a_1^2 + 8})$ so that $\nu \to +0$ for $a_1 \to -\infty$ and $\nu \to +\infty$ for $a_1 \to +\infty$. Equation (4.25) turns to $q_2 = 0.5a_1$ in the case when the speed of diffusive spread is zero, i.e., for $\beta = 0.5$. Here we recall that, for the choice of conditions at infinity as (4.6), $a_1 < 0$ corresponds to migrations enhancing invasion and $a_1 > 0$ corresponds to migrations hampering invasion (= enhancing species retreat).

A point of interest is how much the speed of invasion can be modified by the impact of migration when $\beta < 0.5$. To address this issue, we consider the "speed accretion," i.e.,

$$\Delta q = q_2 - \frac{(2\beta - 1)}{\sqrt{2}} \tag{4.26}$$

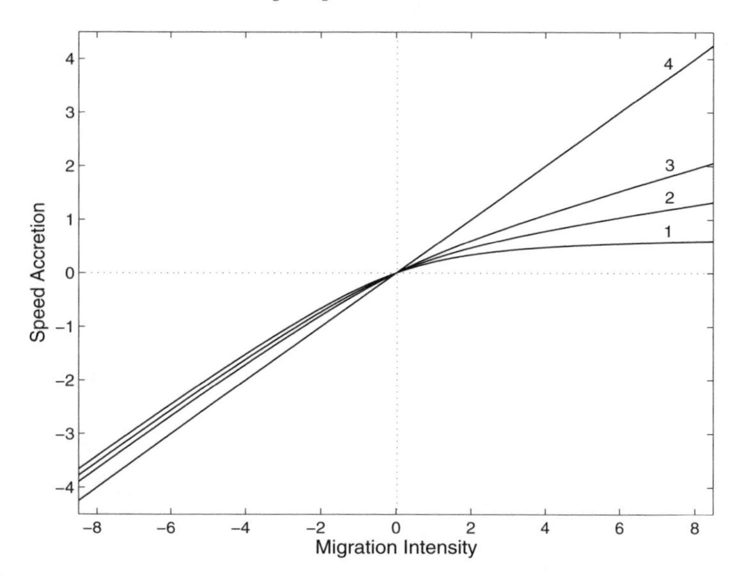

FIGURE 4.1: The speed accretion Δq acquired by the traveling population front as a result of small-scale migrations is shown versus migration intensity a_1 and for different values of the threshold density, curve 1 for $\beta = 0$, curve 2 for $\beta = 0.1$, curve 3 for $\beta = 0.2$ and curve 4 for $\beta = 0.5$.

where the second term in the right-hand side gives the speed of invasion in the nonmigration case, cf. (3.54). The value of Δq versus migration intensity a_1 is shown in Fig. 4.1 for different β, curve 1 for $\beta = 0$, curve 2 for $\beta = 0.1$, curve 3 for $\beta = 0.2$ and curve 4 for $\beta = 0.5$. Negative values of the speed accretion correspond to faster invasion. Note that for each of these curves the speed of diffusive spread has a fixed value, i.e., 0.71, 0.57, 0.42 and 0, respectively. It is readily seen that, already for a_1 being on the order of -1, the speed of the propagating front can be twice as high as it would be without migration.

Since the species spreads into the region with low population density (i.e., from right to left, see (4.6)) when $q_2 < 0$, from relation (4.25) we readily obtain the following condition of successful invasion:

$$\beta < \frac{1}{\nu^2} \, . \tag{4.27}$$

In the case $a_1 = 0$ (no migration), $\nu = \sqrt{2}$ and (4.27) coincides with (3.55). The critical relation $\beta = \nu^{-2}$ is shown in Fig. 4.2 by the solid curve. Thus, the impact of small-scale migrations significantly affects the species' capacity to invasion. In the nonmigration case, the critical magnitude of the Allee effect is given by $\beta = 0.5$ (shown by the dotted line) so that the species is invasive only for $\beta < 0.5$. However, in the case that diffusive spread is supported by migration, the species can remain invasive also for $0.5 < \beta < 1$, cf. the curvilinear triangle in the upper left-hand side of Fig. 4.2.

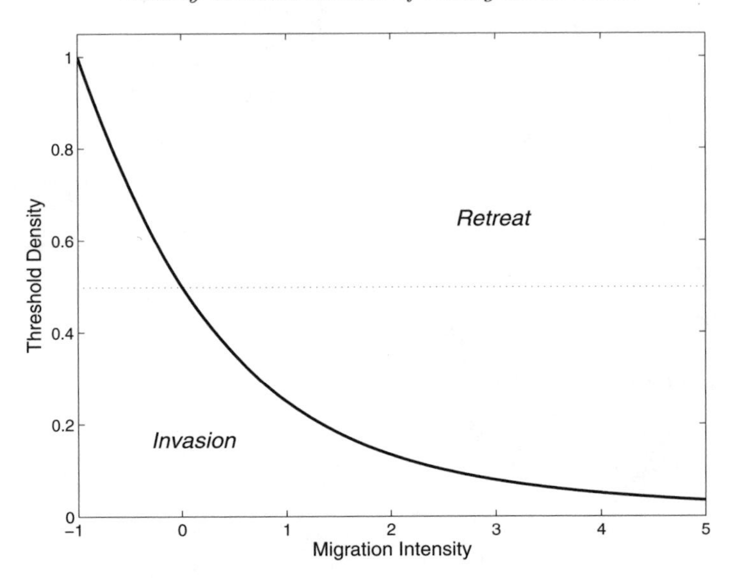

FIGURE 4.2: A map in the parameter plane (a_1, β) of the advection-diffusion-reaction equation (4.12).

Note that, unlike the density-independent case, cf. (4.11), the right-hand side of inequality (4.27) is always positive (which means that the solid curve in Fig. 4.2 never touches the horizontal axis). Consequently, the interplay between the diffusive spread and migration appears to be different. Even a strong "counteractive" migration (corresponding to large positive a_1) cannot block the species invasion caused by the random dispersion of the individuals in case the Allee effect (quantified by parameter β) is sufficiently small.

4.1.3 General case

In a general case, species transport takes place due to both density-dependent and density-independent factors. The dynamics of a given population is then described by full equation (4.5) where now $a_0 \neq 0$ and $a_1 \neq 0$. The results of the two preceding sections immediately apply to this case leading to the following exact solution:

$$u(x,t) = \frac{\beta \exp\{\omega_1[x-(q_1+a_0)t+\psi_{01}]\} + \exp\{\omega_2[x-(q_2+a_0)t+\psi_{02}]\}}{1+\exp\{\omega_1[x-(q_1+a_0)t+\psi_{01}]\} + \exp\{\omega_2[x-(q_2+a_0)t+\psi_{02}]\}}$$

where the notations are the same as in Eq. (4.22). Particularly, in the large-time limit, this solution takes the following form describing propagation of a population front:

$$u(x,t) \simeq \frac{\exp\{\omega_2[x-(q_2+a_0)t+\psi_{02}]\}}{1+\exp\{\omega_2[x-(q_2+a_0)t+\psi_{02}]\}} . \tag{4.28}$$

The condition of successful invasion now takes the form $q_2 < -\ a_0$ which, after a little algebra, reads as follows:

$$\beta < r(\nu,\ a_0) \ = \ \frac{1}{\nu^2} - \frac{a_0}{\nu}\,. \tag{4.29}$$

It is readily seen that $r \geq 0$ for $\nu \leq 1/a_0$ and r is negative otherwise. Thus, since β is assumed to be nonnegative, for any fixed β and ν inequality (4.29) is violated in the case a_0 is positive and sufficiently large. In agreement with the results of Section 4.1.1, it means that the species invasion can always be blocked or reversed in case of sufficiently strong counteractive advection (such as cross-wind or cross-current) provided that the density-dependent migrations are either absent or enhance the species retreat (which corresponds to $a_1 \geq 0$ and $\nu \geq \sqrt{2}$). However, another property of relation (4.29) is that, for any fixed positive value of a_0, however large it can be, inequality (4.29) becomes true for sufficiently small ν. Small ν corresponds to large negative a_1, i.e., to the case when migration takes place towards the region where the species is absent. It means that even strong counteractive advection cannot stave off invasion of given alien species in case its diffusive spread is supported by sufficiently intense small-scale migration.

Thus, the three considered mechanisms of species invasion, i.e., diffusive spread, advection and density-dependent migration, create a certain hierarchy. When considered pairwise, advection appears to have a higher rank than diffusive spread, and diffusive spread has a higher rank than migration. However, the hierarchy is broken if we consider a cooperative impact of different processes: a joint rank of the two "junior" members of hierarchy, i.e., diffusive spread plus migration, appears to be higher than that of advection.

4.2 Accelerating population waves

Population fronts described by the diffusion-reaction equations considered above propagate with a constant speed, although the value of speed can be affected by a variety of factors such as the type of density-dependence in the population growth, impact of other types of species transport, etc. This is congenial to what is typically observed in nature. However, there also can be another scenario of species invasion when the front speed does not remain the same but increases with time; some examples of field observations can be found in the book by Shigesada and Kawasaki (1997). The fronts of that kind are called accelerating or dispersive.

The first attempts that have been made to explain this phenomenon tended to ascribe it to the impact of the "non-Gaussian" diffusion which can manifest itself either via fat-tailed dispersive kernels in the corresponding integral-difference equations (Kot et al., 1996) or via scale-dependence of the diffusion coefficient in the diffusion-reaction equations (Petrovskii, 1999b). Somewhat later, however, it was shown that accelerating waves can arise as well in a system with usual Gaussian diffusion as a result of prolonged impact of the initial conditions (Petrovskii and Shigesada, 2001). Remarkably, the latter study was done based on an exactly solvable model of population dynamics.

We assume that the alien population spreads in a homogeneous environment and the initial species distribution exhibits radial symmetry. Then it can be expected that the population density depends only on the distance from the origin and the dynamics of the invading species is described by the following equation:

$$\frac{\partial u}{\partial t} = D\left(\frac{\partial^2 u}{\partial r^2} + \frac{\eta}{r}\frac{\partial u}{\partial r}\right) + F(u) \tag{4.30}$$

where $r = (x^2 + y^2)^{1/2}$, $0 \le r < L$ and L is the radius of the overall area accessible for invasion. In the most ecologically relevant case of cylindrical symmetry $\eta = 1$; in a more general stuation, coefficient η can be equal to 0, 1 or 2 in the cases of planar, cylindrical and spherical symmetry, respectively.

Eq. (4.30) should be provided with the initial distribution of the species, $u(r, 0)$. The type of the initial condition depends on the nature of the problem under consideration. Concerning the problem of biological invasion, the invasive species first appears in a small domain inside a given area. Mathematically, it means that the function $u(r, 0)$ should be chosen either finite or "localized," i.e., promptly decreasing to zero with an increase in the distance from the center of the domain. In both cases, the behavior of the function $u(r, t)$ can be characterized by the typical radius l of the domain:

$$u(r, 0) = \Phi(r;\ l) \tag{4.31}$$

where function Φ promptly approaches zero when $r/l \gg 1$.

A proper choice of the dimensionless variables, i.e., in our case, the choice of scales for the variables u, r and t, is an important point. As usual, we assume that function $F(u)$ allows for at least two stationary homogeneous states, i.e., $F(0) = F(K) = 0$, parameter K being the carrying capacity for the given population. Since K serves as a natural scale for the species concentration u, it is convenient to consider $\tilde{u} = u/K$. Then we can write $F(u) = (K/\tau)\tilde{F}(\tilde{u})$ where parameter τ has the dimension of time and $\tilde{F}(\tilde{u})$ is now dimensionless, $\tilde{F}(0) = \tilde{F}(1) = 0$. Thus, from Eq. (4.30) we obtain:

$$\frac{\partial \tilde{u}}{\partial t} = D\left(\frac{\partial^2 \tilde{u}}{\partial r^2} + \frac{\eta}{r}\frac{\partial \tilde{u}}{\partial r}\right) + \frac{1}{\tau}\tilde{F}(\tilde{u})\ . \tag{4.32}$$

Including the scaling factor $1/\tau$ explicitly into Eq. (4.32) implies that function $\tilde{F}(\tilde{u})$ is normalized, i.e., satisfies certain conditions. Generally, various biologically relevant constraints can be used. Here we suggest that $d\tilde{F}(\tilde{u} = 0)/d\tilde{u} = 1$.

The choice of the scale for position r is, to a large extent, determined by the goals of the modeling. Generally, in the problem of the evolution of a finite initial perturbation there are a few different length-scales: the (typical) radius L of the whole domain, the typical radius l of the perturbation and the "diffusive length" $(D\tau)^{1/2}$. If one is interested in the large-time limit of the system dynamics, when the perturbation has spread over the whole area and the impact of the boundaries cannot be neglected, then the length L seems to be a natural scale. In case of smaller but still large times when the effect of the boundaries is not yet essential but the transients caused by particulars of the initial conditions have already died out, cf. "intermediate asymptotics" (Barenblatt and Zeldovich, 1971), the diffusive length is the most reasonable choice. However, in case the early stages of the system evolution are of primary interest when the influence of the initial conditions has not yet completely disappeared, it seems that the characteristic length l of the initial perturbation also can be taken as a suitable length-scale.

For reasons that will become clear later, here we are mainly concerned with the early stages of the system dynamics. Therefore, we consider $\tilde{r} = rl^{-1}$ and $\tilde{t} = tDl^{-2}$. Then we arrive at the following problem containing now only dimensionless values (tildes will be omitted hereafter):

$$\frac{\partial u}{\partial t} = \left(\frac{\partial^2 u}{\partial r^2} + \frac{\eta}{r}\frac{\partial u}{\partial r}\right) + f(u) \tag{4.33}$$

and

$$u(r,0) = \Phi(r;\ 1) \tag{4.34}$$

where $f(u) = (1/\gamma)F(u)$ and $\gamma = D\tau l^{-2}$.

As an immediate generalization of logistic growth rate, we consider the function $f(u)$ belonging to the following family:

$$f_\gamma(u) = \frac{1}{\gamma}\, u\,(1 - u^\gamma)\,. \tag{4.35}$$

Let us note that, for every particular case the value of the parameter γ is fixed. However, in order to be able to compare different cases of invasion in terms of Eq. (4.33), we consider the behavior of the solutions of the problem (4.33–4.34) assuming that γ can take any positive value.

The family (4.35) has small-γ limit:

$$\lim_{\gamma\to 0} f_\gamma(u) = u\ \lim_{\gamma\to 0}\left[\frac{1}{\gamma}(1 - u^\gamma)\right] = \bar{f}(u) \tag{4.36}$$

where

$$\bar{f}(u) = -\,u\ln u \tag{4.37}$$

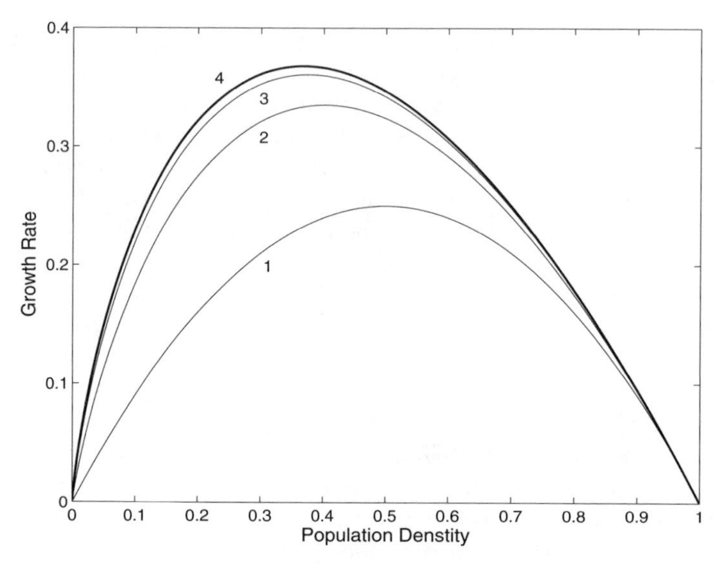

FIGURE 4.3: Population growth rate $f_\gamma(u)$ as given by (4.35) for $\gamma = 1$ (curve 1), $\gamma = 0.2$ (curve 2) and $\gamma = 0.04$ (curve 3); curve 4 shows the small-γ asymptotics $\bar{f}(u)$, cf. (4.37) (with permission from Petrovskii and Shigesada, 2001).

where $\ln u \equiv \log_e u$, $e = 2.718\ldots$. Note that, for all positive γ and $u \geq 0$ (letting $\bar{f}(u) = 0$ for $u = 0$),

$$f_\gamma(u) \leq \bar{f}(u) \ . \tag{4.38}$$

Simple relations (4.36–4.38) will appear to have an important meaning for further consideration.

A few functions of the family (4.35), as well as asymptotics (4.37), are shown in Fig. 4.3; curve 1 for $\gamma = 1$, curve 2 for $\gamma = 0.2$, curve 3 for $\gamma = 0.04$ and curve 4 for $\bar{f}(u)$. One can see that, while for the values of γ on the order of unity there is a significant discrepancy between $f_\gamma(u)$ and $\bar{f}(u)$, for γ smaller than about 0.1 the plots of $f_\gamma(u)$ and $\bar{f}(u)$ lie very close to each other. Thus, although the biological meaning of $\bar{f}(u)$ may seem to be somewhat obscure because it implies an infinite per capita growth rate when the species density u tends to zero, it provides a good approximation for the members of the family (4.35). The smaller γ is the higher is the accuracy.

Investigation of the problem (4.33–4.34) will be performed based on an exact solution of Eq. (4.33), which we introduce in the next section. Let us note here that Eq. (4.33) is nonlinear and that makes it virtually impossible to obtain its general solution, i.e., for an arbitrary $u(r, 0)$. To be able to solve Eq. (4.33) analytically, we have to make particular suggestions concerning the form of the initial conditions. Here we assume that

$$u(r,0) = g_0 \, \exp\left[-\left(\frac{r}{R_{in}} \right)^2 \right] \tag{4.39}$$

where R_{in} and g_0 are positive dimensionless parameters. Although function (4.39) is not finite, we assume that its very fast decay for $r \to \infty$ makes it appropriate for description of the invasive species initial distribution, cf. Section 7.2.

Let us note that, at first glance, parameter R_{in} in Eq. (4.39) may seem redundant because variable r is already scaled with the typical radius of the initially invaded domain. In fact, this is not so. The problem is that the definition of a "typical" radius usually involves uncertainty. For instance, the initial radius l can be defined as the radius of the domain where the species density exceeds a certain prescribed value u_*. However, in this case, l would depend essentially on u_* which, in its turn, may be known only approximately. Thus it seems important to understand how the solution's behavior can change with respect to variation of the value of the typical radius.

Another point is that our choice of dimensionless variables when $\tilde{r} = r/l$ is not the only possible one. Considering other scales for r and t (see the paragraph below Eq. (4.32)), we would immediately arrive at the equations similar to (4.33)–(4.34) but with different coefficients. Thus, in order to retain mathematical generality in further considerations, we assume that the value of R_{in} can be arbitrary.

4.2.1 Self-similar exact solution

Now we proceed to derive an exact solution to Eqs. (4.33–4.34). Let us try to find a solution in the following special self-similar form:

$$u(r,t) = g(t)\phi(\xi) \quad \text{where} \quad \xi = \frac{r}{R(t)} \; , \tag{4.40}$$

where functions $g(t)$ and $R(t)$ are to be determined while $\phi(\xi)$ is fixed by the choice of the initial conditions. The idea of further analysis is that a suitable choice of the form of initial conditions, i.e., of function $\phi(\xi)$, can make Eq. (4.33) solvable. Thus, Eq. (4.39) works as an ansatz.

Having substituted Eqs. (4.40) into Eq. (4.33), we arrive at the following equation:

$$\frac{g}{R^2}\frac{d^2\phi}{d\xi^2} + \left(\frac{g\xi}{R}\frac{dR}{dt} + \frac{\eta}{\xi}\frac{g}{R^2} \right)\frac{d\phi}{d\xi} - \phi\frac{dg}{dt} + f(g\phi) = 0 \; . \tag{4.41}$$

Briefly, the method is as follows. Eq. (4.41) contains two unknown functions, g and R. A suitable choice of function ϕ after substitution into Eq. (4.41) "splits" the equation into two parts showing different functional dependence on ξ (e.g., containing different powers of ξ). When structured in this way, the equation is equivalent to the system of two equations which can be used to obtain $g(t)$ and $R(t)$.

Restricting our consideration to the case $\phi(\xi) = e^{-\xi^2}$ (consistent with the choice of initial condition in the form (4.39)), from Eq. (4.41) we obtain

$$\frac{g}{R^2}\left(4\xi^2 - 2\right)e^{-\xi^2} - 2\left(\frac{g\xi^2}{R}\frac{dR}{dt} + \frac{\eta g}{R^2}\right)e^{-\xi^2} \qquad (4.42)$$
$$- \frac{dg}{dt}e^{-\xi^2} + f\left(ge^{-\xi^2}\right) = 0.$$

Obviously, applicability of the approach depends on the form of the function f. For biological reasons it would be interesting to consider a case when $f(u)$ belongs to the family (4.35). It is not difficult to recognize, however, that for any $f_\gamma(u)$ Eq. (4.42) has only the trivial solution $g(t) \equiv 0$.

A nontrivial solution can be obtained if we do not strictly stick to immediate biological reasons. Namely, let us consider $f(u) = \bar{f}(u)$; see Eq. (4.37). Having substituted it into Eq. (4.42), after simple transformations we obtain the following equation:

$$\left(4\,\frac{g}{R^2} - 2\,\frac{g}{R}\frac{dR}{dt} + g\right)\xi^2 \qquad (4.43)$$
$$- \left(2\,\frac{g}{R^2} + 2\,\frac{\eta g}{R^2} + \frac{dg}{dt} + g\ln g\right) = 0.$$

Since ξ^2 and 1 are linearly independent, the coefficients of both must be identically zero, so that from Eq. (4.43) we arrive at the following system for the two unknown functions g and R:

$$\frac{dR}{dt} = \frac{R}{2} + \frac{2}{R}\,, \qquad (4.44)$$

$$\frac{d\ln g}{dt} = -\ln g - \frac{\sigma}{R^2} \qquad (4.45)$$

where $\sigma = 2 + 2\eta$.

Eqs. (4.44–4.45) are easily solved. Indeed, immediate integration of Eq. (4.44) yields the following expression for R:

$$R(t) = \left[\left(4 + R_{in}^2\right)e^t - 4\,\right]^{1/2}\,, \qquad (4.46)$$

where $R(t=0) = R_{in}$. Then, taking Eq. (4.45) together with Eq. (4.46), after integration we obtain this expression for function g:

$$g(t) = \exp\left(e^{-t}\ln g_0 - e^{-t}\frac{2\sigma}{4 + R_{in}^2}\ln\left[\frac{R(t)}{R_{in}}\right]\right) \qquad (4.47)$$

where $g_0 = g(0)$. Thus, Eqs. (4.46), (4.47), taken together with (4.40), give an exact self-similar solution for the problem (4.33), (4.37), (4.39). Evidently,

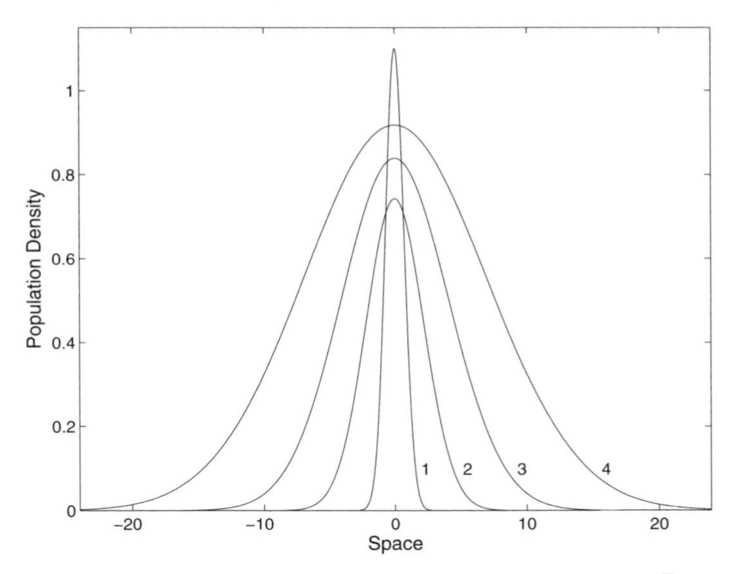

FIGURE 4.4: The exact solution of Eq. (4.33) with $f(u) = \bar{f}(u)$ and the initial condition (4.39). Parameters and further details are given in the text.

function $g(t)$ gives the species concentration at the origin which, according to our choice of function ϕ, coincides with the maximum population density, or "amplitude," of the species' spatial distribution at every moment t. The value of $R(t)$ may be regarded as the radius of the region inhabited by the invading species (see details below). Let us note that the solution (4.46–4.47) describes 1-D, 2-D and 3-D cases depending on the value of coefficient σ.

The exact solution (4.40), (4.46–4.47) for the problem (4.33), (4.39) with $f(u) = \bar{f}(u)$ is shown in Fig. 4.4 for parameters $\sigma = 2$, $R_{in} = 1$ and $g_0 = 1.1$ for $t = 0$ (curve 1), $t = 1$ (curve 2), $t = 2$ (curve 3), $t = 3$ (curve 4) and $t = 4$ (curve 5). Thus one can see that, as it is prescribed by Eq. (4.40), the evolution of the species' spatial distribution does not alter its form; the temporal changes result only in changing the values of the amplitude g of the distribution and the effective radius R of the inhabited domain.

While the behavior of function $R(t)$ is rather simple, the evolution of the amplitude g is somewhat more complicated, being determined by three parameters R_{in}, g_0 and σ. The changes in the behavior of $g(t)$ occurring with variations of parameter values are illustrated in Figs. 4.5 to 4.7. Fig. 4.5 shows function $g(t)$ for $\sigma = 2$ and $R_{in} = 1$ and different values of the initial amplitude $g_0 = 2.5$ (curve 1), $g_0 = 1.0$ (curve 2), $g_0 = 0.25$ (curve 3) and $g_0 = 0.1$ (curve 4). Fig. 4.6 exhibits function $g(t)$ for $\sigma = 2$, $g_0 = 1$ and for different values of initial radius $R_{in} = 2.0$ (curve 1), $R_{in} = 1.0$ (curve 2), $R_{in} = 0.4$ (curve 3) and $R_{in} = 0.1$ (curve 4). Fig. 4.7 shows $g(t)$ for $g_0 = R_{in} = 1$ for $\sigma = 2$ (curve 1), $\sigma = 4$ (curve 2) and $\sigma = 6$ (curve 3), i.e., for planar, cylindrical and spherical cases, respectively. Thus one can see that, at least for not

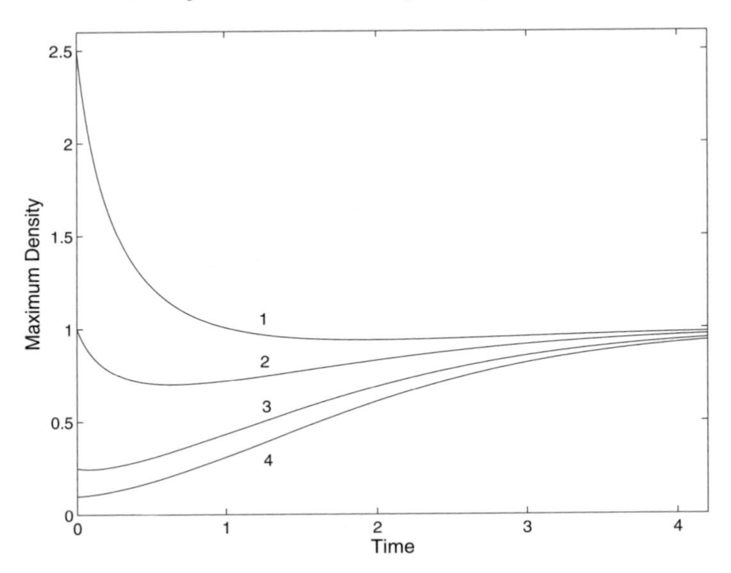

FIGURE 4.5: The amplitude g of the exact solution (4.40), (4.46) and (4.47) versus time shown for $\sigma = 2$, $R_{in} = 1$ and different g_0, curve 1 for $g_0 = 2.5$, curve 2 for $g_0 = 1$, curve 3 for $g_0 = 0.25$, curve 4 for $g_0 = 0.1$.

very large values of initial radius R_{in} or initial amplitude g_0, amplitude $g(t)$ typically shows nonmonotonic behavior, first falling down to its minimal value and then gradually increasing to the asymptotical value $g = 1$ prescribed by Eq. (4.33).

An important point is to understand the relation between the exact solution (4.46–4.47) of Eq. (4.33) obtained for $f(u) = \bar{f}(u)$ and the solutions corresponding to a more biologically reasonable case when the nonlinear term in Eq. (4.33) is described by function $f_\gamma(u)$; see (4.35). Taking into account relation (4.38) and applying the comparison principle for the solutions of nonlinear parabolic equations (see Section 7.4), it is readily seen that

$$u_\gamma(r,t) \le u(r,t) = g(t)\phi\left(\frac{r}{R(t)}\right) \quad \text{for any} \quad r, t \ge 0 \tag{4.48}$$

where u_γ denotes the solution of Eq. (4.33) with $f(u) = f_\gamma(u)$ and the same initial condition (4.39). Thus, functions (4.46) and (4.47) provide an upper bound for, respectively, the radius and the amplitude of the spatial distribution of a population with the growth rate described by (4.35).

Relation (4.48), however, leaves open the question how close the solutions u and u_γ actually can be for different parameter values. In particular, it remains unclear whether the radius of the invaded area as given by (4.46) can be used to predict the rate of species invasion in the biologically meaningful case $f(u) = f_\gamma(u)$. While the behavior of the amplitude $g(t)$ shown by Figs. 4.5 to 4.7 looks reasonable, an exponential growth of radius $R(t)$ is not realistic.

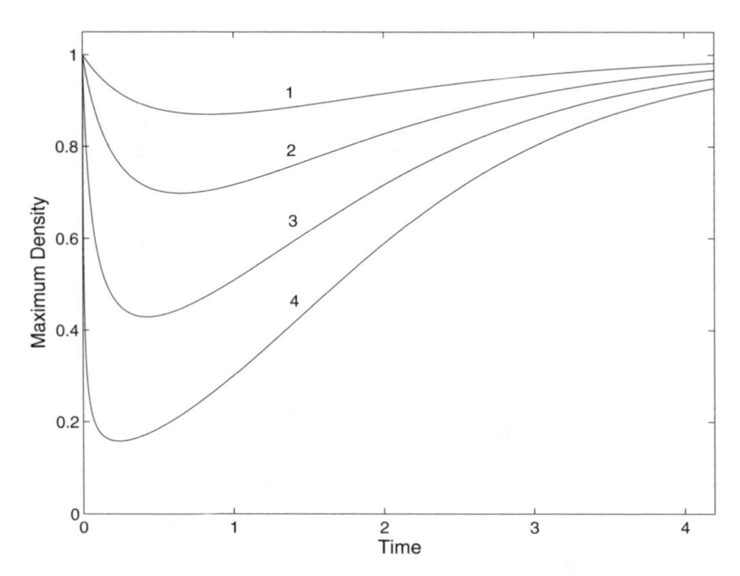

FIGURE 4.6: The amplitude $g(t)$ of the exact solution shown for $\sigma = 2$, $g_0 = 1$ and different R_{in}: curve 1 for $R_{in} = 2$, curve 2 for $R_{in} = 1$, curve 3 for $R_{in} = 0.4$ and curve 4 for $R_{in} = 0.1$.

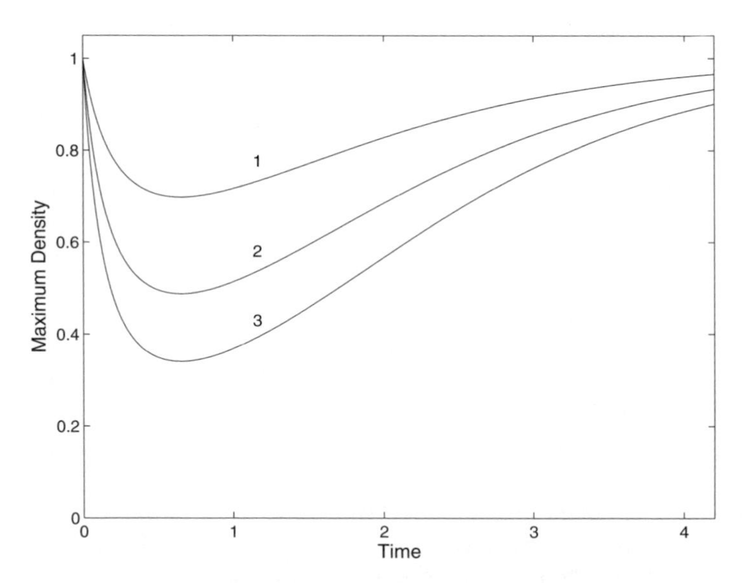

FIGURE 4.7: The amplitude $g(t)$ of the exact solution shown for $R_{in} = g_0 = 1$ and different dimensionality of the problem, curve 1 for 1-D case ($\sigma = 2$), curve 2 for 2-D case ($\sigma = 4$) and curve 3 for 3-D case ($\sigma = 6$).

Rigorous mathematical analysis (Kolmogorov et al., 1937) shows that, for sufficiently large times, a population with the local growth rate described by function $f_\gamma(u)$ with an arbitrary $\gamma > 0$ spreads over the area via propagation of a stationary (= with constant speed and shape) traveling population wave; it means that $R(t)$ must increase linearly. On the other hand, taking into account that $\bar{f}(u)$ provides a good estimate for function $f_\gamma(u)$ for sufficiently small values of γ, one can expect that solution (4.46) may give a good approximation for the "realistic" solution at the stage preceding the stationary wave propagation.

In order to address this issue, we are going to compare the exact solution (4.40), (4.46–4.47) with the results of numerical integration of Eq. (4.33) obtained for $f(u) = f_\gamma(u)$ and the same initial condition (4.39).

Aiming to make the comparison more straightforward, here we admit a definition of the radius of the inhabited (invaded) area that is different from the standard one. Namely, according to the standard definition, cf. Shigesada and Kawasaki (1997), the radius r_* of invaded area at every moment t is given by the following equation:

$$u(r_*, t) = u_* \tag{4.49}$$

where u_* is a certain threshold density so that the species can hardly be detected when $u \leq u_*$. Alternatively, however, one can define the radius $\bar{r}$ of the inhabited domain as a distance where the species density becomes M times smaller than its maximal value, i.e.,

$$\frac{u(\bar{r}, t)}{u(0, t)} = \frac{1}{M} \tag{4.50}$$

suggesting that the maximum of the species density is reached in the center of the domain as it is in the case considered above. Here $M > 1$ is a parameter. The latter definition seems to be convenient in the special case of self-similar spreading of the population, cf. Eq. (4.40). It is readily seen that, in this case, the area inside the circle of radius $\bar{r}(t)$ at every moment contains a fixed fraction of the whole population. Particularly, for exact solution (4.40), (4.46–4.47), $\bar{r}(t) = R(t)$ if we set $M = e$. For an arbitrary value of M, $\bar{r}(t) = R(t)\sqrt{\ln M}$. Note that, if $u(0, t)$ tends to unity when t goes to infinity (as it takes place for the solutions of the problem (4.33)–(4.35)), the definitions (4.49) and (4.50) are apparently consistent for large time (considering $M = 1/u_*$).

It must be mentioned that, for an early stage of the invasion, the behavior of r_* and $\bar{r}(t)$ (or $R(t)$ for a special case) can be different. Applying the standard definition (4.49) to the exact self-similar solution given by Eqs. (4.40), (4.46–4.47), we obtain:

$$u(r_*, t) = g(t) \exp\left[-\left(\frac{r_*}{R(t)}\right)^2\right] = u_* \tag{4.51}$$

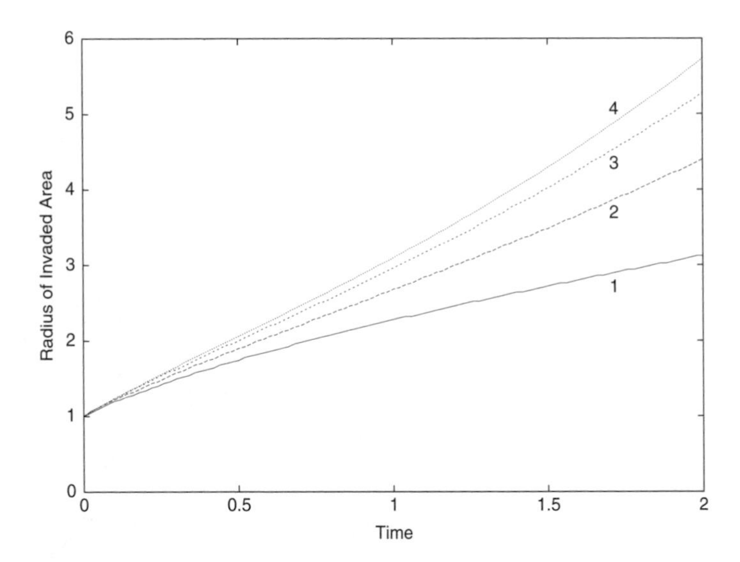

FIGURE 4.8: The radius of invaded area obtained by numerical integration of Eq. (4.33) with $f(u) = f_\gamma(u)$ for $\gamma = 1$ (curve 1), $\gamma = 0.2$ (curve 2) and $\gamma = 0.04$ (curve 3); curve 4 shows the exact solution (4.46) (with permission from Petrovskii and Shigesada, 2001).

so that

$$r_*(t) = R(t)\left[\ln\frac{g(t)}{u_*}\right]^{1/2}. \tag{4.52}$$

The radius $\bar{r}(t)$ defined according to Eq. (4.50) is given by the monotonically increasing function (4.46) for any values of parameters R_{in} and M. The behavior of r_* calculated as (4.52) depends on the relation between the values of u_*, R_{in} and g_0. Particularly, since function $g(t)$ can be nonmonotonic (see Figs. 4.5 to 4.7), the behavior of r_* can be also nonmonotonic at early stages of the process.

Now, to run numerical simulations, we restrict ourselves to the 1-D case. Let us begin with the radius $R(t)$ of the invaded domain. Here and below, the radius R of the invaded domain is determined according to Eq. (4.50) with $M = e$. Fig. 4.8 shows the radius R versus time obtained numerically for $\gamma = 1$ (curve 1), $\gamma = 0.2$ (curve 2) and $\gamma = 0.04$ (curve 3) as well as the exact solution (4.46) (curve 4). It is readily seen that, for a certain period of time at the beginning of the system evolution, the behavior of $R(t)$ given by (4.46) is practically indistinguishable from the results obtained for a population with generalized logistic growth. The smaller γ is the longer is the period of closeness between the solutions. This relation has a clear biological meaning because parameter γ is proportional to the reproduction time τ of the invasive species; see the line below Eq. (4.34). Thus, the higher is the species' ability for reproduction, the more applicable is the exact solution (4.40), (4.46–4.47) to describe the species invasion.

Contrary to $R(t)$, the amplitude $g(t)$ of the species concentration provided by Eq. (4.47) is in a very good agreement with the results obtained for the case $f(u) = f_\gamma(u)$ for large t; at earlier stages, the relation between the solutions essentially depends on γ. Fig. 4.9 shows amplitude g versus time calculated numerically for $R_{in} = 1$ for (a) $g_0 = 0.1$ and (b) $g_0 = 1.2$ for $\gamma = 1$ (curves 1) and $\gamma = 0.2$ (curves 2); curves 3 show $g(t)$ given by (4.47). Thus, whereas for γ on the order of unity there may be a considerable discrepancy between the exact and numerical solutions at the beginning of the process, cf. Fig. 4.9a, for the values of γ less than about 0.2, Eq. (4.47) gives a good approximation for the amplitude obtained in the "realistic" case $f(u) = f_\gamma(u)$ for any moment of time.

It should be noted that, while in an unbounded domain the early stages of the system evolution may probably be regarded as insignificant because of their relatively short duration, the situation can be different in the case of a domain of finite length. In such a case it may appear that the population spread will never reach its constant-rate asymptote. For example, Fig. 4.10 shows the radius of the inhabited domain calculated numerically in the 1-D case (curve 1) for parameters $\gamma = 0.04$, $R_{in} = 1$, $g_0 = 1.2$ and the radius of the overall domain $L = 30$. One can see that just at the time when the system nearly reaches its traveling wave regime (i.e., for $t \approx 6$), the impact of the boundaries begins to prevail and the radius of the inhabited domain promptly approaches its stationary asymptote $R \simeq L = 30$. In this case the total duration of "nontrivial" dynamics $t \simeq 6$ while the asymptote (4.46) (curve 2) gives a reasonable approximation for $R(t)$ for $t \leq 4$, i.e., a value of the same order. Let us also mention that, in this example, the traveling wave asymptote is hardly applicable to any period of the system evolution. Although for the values of t between 5 and 6 the plot of $R(t)$ (curve 1 in Fig. 4.10) nearly looks as a straight line, a closer inspection shows that its slope is still about 1.5 times smaller than it is predicted to be in case of the traveling wave propagation, see Eq. (2.19).

We want to emphasize that Figs. 4.8 and 4.10 show the radius of invaded area in dimensionless units. After being re-scaled back to original dimensional units, $t = 1$ may correspond to as much as several months, or even years. Indeed, taking $D = 1\ \text{km}^2/\text{year}^{-1}$ (which is consistent with diffusivity of some insect species) and assuming that the area inhabited prior to the beginning of geographical spread is $1\ \text{km}^2$, from the definition of dimensionless variables we obtain $l^2/D = 1$ year. A more specific example will be considered in Chapter 8.

In conclusion, we want to check whether the scenario of the accelerating species spread considered above is robust, at least, qualitatively, to the definition of the radius of invaded area, cf. the standard definition (4.49) and the modified definition (4.50). For this purpose, Fig. 4.10 also shows the plots of $r_*(t)$ obtained for $u_* = 0.02$ (curve 3) and $u_* = 0.1$ (curve 4); other parame-

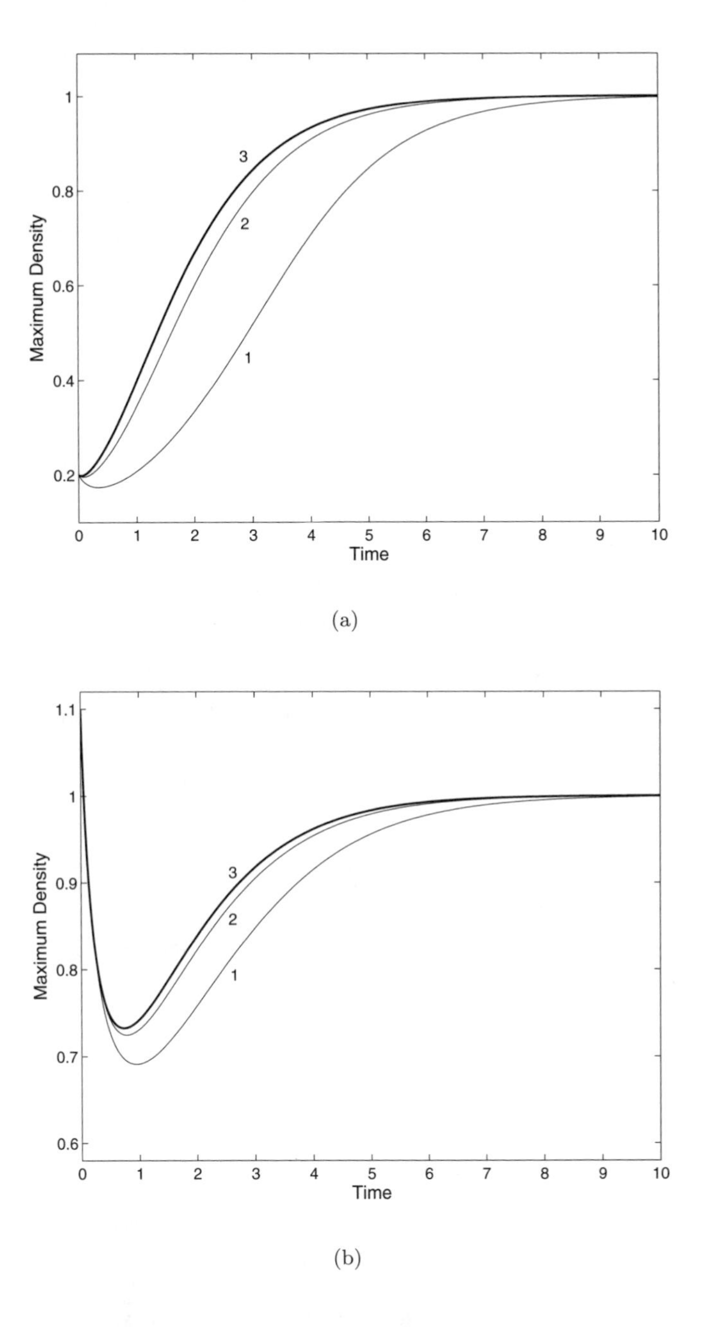

FIGURE 4.9: The maximum value of the population density obtained by numerical integration of Eq. (4.33) with $f(u) = f_\gamma(u)$ for $\gamma = 1$ (curves 1) and $\gamma = 0.2$ (curves 2); (a) and (b) correspond to $g_0 = 0.1$ and $g_0 = 1.2$, respectively. Curves 3 show the exact solution (4.47).

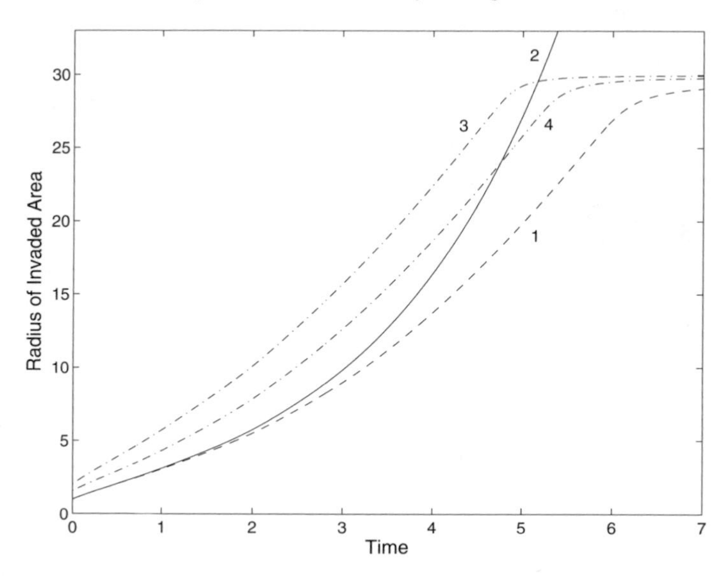

FIGURE 4.10: The radius of invaded area calculated in the case that species spread takes place in a bounded domain. Curve 1 shows the radius versus time obtained by numerical integration of Eq. (4.33) with $f(u) = f_\gamma(u)$ for parameters $\gamma = 0.04$, $R_{in} = 1$ and $g_0 = 1.2$; curve 2 shows the exact solution (4.46). Curves 3 and 4 show the radius calculated according to the alternative definition (4.51) for $u_* = 0.02$ and $u_* = 0.1$, respectively (with permission from Petrovskii and Shigesada, 2001).

ters are the same. Thus, although in general r_* can exhibit more complicated behavior (e.g., for higher values of u_* function $r_*(t)$ can become nonmonotonic), both functions possess similar properties; particularly, both $R(t)$ and r_* show apparent initial acceleration in the speed of invasion.

4.3 The problem of critical aggregation

Biological invasion typically begins with a local event when a number of organisms of an alien species are introduced into a given ecosystem. Under favorable conditions, the new population may begin to grow and spread onto new areas which may eventually lead to its geographical spread, either via population front propagation or in a more complicated manner. Not every introduction of a new species leads to its geographical spread. The outcome of the species introduction depends on a variety of factors such as environmental conditions, availability of food or nutrients, presence/absence of natural enemies, species evolutionary traits, etc. It also depends on the initial population size of the introduced species so that the larger the size is the higher are the chances for the species' "success." The problem of distinguishing between the cases when a local invasion leads to extinction of the introduced alien species and when it leads to its geographical spread is called the problem of critical aggregation.

A closer consideration of the latter issue, however, shows that this problem is not so simple and the "success" of the introduction depends not just on the initial population size but more on the shape and radius of the infested area and the maximum population density inside. Indeed, it is readily accepted that low population density is likely to lead to species extinction, e.g., as a result of environmental/demographical stochasticity or due to the Allee effect (May, 1972; Gilpin, 1972; Lewis and Kareiva, 1993; Courchamp et al., 1999). Therefore, a large population size may result in species extinction in case the population is spread over a large area so that the population density appears to be very small. On the contrary, a smaller population size may result in species persistence/invasion in case the population is concentrated in a sufficiently small area.

Probably the first mathematical consideration of the problem of critical aggregation was done by Kierstead and Slobodkin (1953) in application to the dynamics of plankton patches in marine ecosystems. It has been observed in many field studies that large patches and small patches behave in a different way: while small patches tend to disappear, large patches tend to grow. Kierstead and Slobodkin assumed that this behavior arises as a result of the interplay between plankton multiplication and marine turbulence and considered the following simple model:

$$u_t(x,t) = Du_{xx} + \alpha u \tag{4.53}$$

where α is the plankton growth rate and D is turbulent diffusivity. Here $0 < x < L$ where L is the radius of the patch, and the environmental conditions outside of the patch are assumed to be unfavorable for plankton growth so that

$$u(0,t) \;=\; 0, \quad u(L,t) \;=\; 0 \tag{4.54}$$

and $u(x,t) \equiv 0$ for $x < 0$ and $x > L$.

The solution of problem (4.53–4.54) can be easily found applying the standard method of variables separation. Let us look for a solution of Eq. (4.53) in the following form:

$$u(x,t) = \psi(t)\phi(x) \tag{4.55}$$

where ψ and ϕ are certain functions to be determined. Having substituted it to (4.53), we obtain:

$$\frac{1}{\psi}\left(\frac{d\psi}{dt} - \alpha\psi\right) \;=\; \frac{D}{\phi}\frac{d^2\phi}{dx^2} \;=\; -\,\mu \tag{4.56}$$

where μ is a certain constant: indeed, since the left-hand side of Eq. (4.56) depends on t but not on x, and the right-hand side depends on x but not on t, they only can be constant.

From (4.56), we arrive at

$$D\frac{d^2\phi}{dx^2} + \mu\phi = 0, \tag{4.57}$$

$$\frac{d\psi}{dt} + (\mu - \alpha)\psi = 0. \tag{4.58}$$

Note that, due to the boundary conditions (4.54), $\phi(0) = \phi(L) = 0$. It is straightforward to see that Eq. (4.57) does not have nontrivial solutions satisfying these conditions for $\mu \leq 0$. In case $\mu > 0$, the solution of Eq. (4.57) is

$$\phi(x) = A\cos\left(x\sqrt{\frac{\mu}{D}}\right) + B\sin\left(x\sqrt{\frac{\mu}{D}}\right), \tag{4.59}$$

A and B are to be found. From $\phi(0) = 0$ we obtain that $A = 0$, and from $\phi(L) = 0$ we obtain (assuming that $B \neq 0$) that $\sin(L\sqrt{\mu/D}) = 0$, so that

$$\mu = \mu_n = \left(\frac{\pi n}{L}\right)^2 D, \quad n = 1, 2, \ldots . \tag{4.60}$$

Thus, for any $\mu = \mu_n$,

$$\phi(x) = \phi_n(x) = B_n \sin\left(\frac{\pi n x}{L}\right). \tag{4.61}$$

Correspondingly, the solution of Eq. (4.57) is

$$\psi_n(t) = C_n e^{(\alpha - \mu_n)t}. \tag{4.62}$$

The product $\psi_n(t)\phi_n(x)$ gives a partial solution of (4.53). Since Eq. (4.53) is linear, the sum of any of its partial solutions is also a solution. Therefore, the general solution of (4.53) allowing for the boundary conditions (4.54) is:

$$u(x,t) = \sum_{n=1}^{\infty} \tilde{C}_n e^{(\alpha-\mu_n)t} \sin\left(\frac{\pi n x}{L}\right) \tag{4.63}$$

where the coefficients $\tilde{C}_n$ are determined by the initial conditions.

Remarkably, the behavior of the solution (4.63) is different for different parameters α, D and L. Indeed, it is readily seen that for $\alpha < \mu_1$ it decays with time while for $\alpha > \mu_1$ it exhibits an unbounded growth. The critical radius of the patch is given by the equation $\alpha = \mu_1$ so that

$$L_{cr} = \pi\sqrt{\frac{D}{\alpha}} . \tag{4.64}$$

The above considerations are immediately extended onto 2-D and 3-D cases with cylindrical and spherical symmetry, respectively. The result remains essentially the same, the only difference is a numerical coefficient on the order of unity in Eq. (4.64). The unbounded growth predicted by the solution (4.63) if $L > L_{cr}$ is evidently an artifact of the assumption that the per capita population growth α is not density-dependent; from the standpoint of real populations it means that the population density should be small enough.

Although Eq. (4.64) was originally obtained for the dynamics of plankton patches, since Eq. (4.53) is a general one (up to the assumption of the linear population growth), it immediately applies to the problem of critical aggregation. Thus, according to (4.64) the alien population is viable for $L > L_{cr}$ and it is prone to extinction for $L < L_{cr}$ where L is the radius of the initially inhabited area. Note that, in spite of extreme simplicity of the model (4.53), the dependence of L_{cr} on the species diffusivity and the per capita population growth looks biologically reasonable: the higher is the growth rate of the introduced population the smaller the initially inhabited area can be. However, contrary to intuitive expectations, the critical radius given by (4.64) does not depend on the initial population density. As well as the unbounded growth of the population density, this is not an intrinsic property of the critical aggregation but just a consequence of the linear population growth. The dependence of L_{cr} on the initial population density can be revealed by more elaborate models; some of them are considered below.

4.3.1 Practical stability concept

An invaluable contribution from the Kierstead–Slobodkin model was that it was the first to demonstrate the existence of the critical radius in the problem of critical aggregation. However, due to its simplicity, there are many features that it fails to catch. A more realistic approach to the problem is unlikely possible without taking into account the density-dependence of the population growth. In order to provide a more sound consideration, now we are going to make use of the following single-species model allowing for the nonlinear growth:

$$\frac{\partial u}{\partial t} = \left(\frac{\partial^2 u}{\partial r^2} + \frac{\eta}{r}\frac{\partial u}{\partial r}\right) + f(u) \tag{4.65}$$

(in dimensionless variables) where $\eta = 0, 1, 2$ corresponds 1-D, 2-D and 3-D cases, respectively. The initial species distribution is given as

$$u(r, 0) = \Phi(r) \tag{4.66}$$

where $\Phi(r)$ is a certain function.

It should be mentioned that, apart from its density-independence, the Kierstead–Slobodkin model has another serious drawback. The point is that, in the model (4.53–4.54), the radius L of the domain is fixed. Whether the

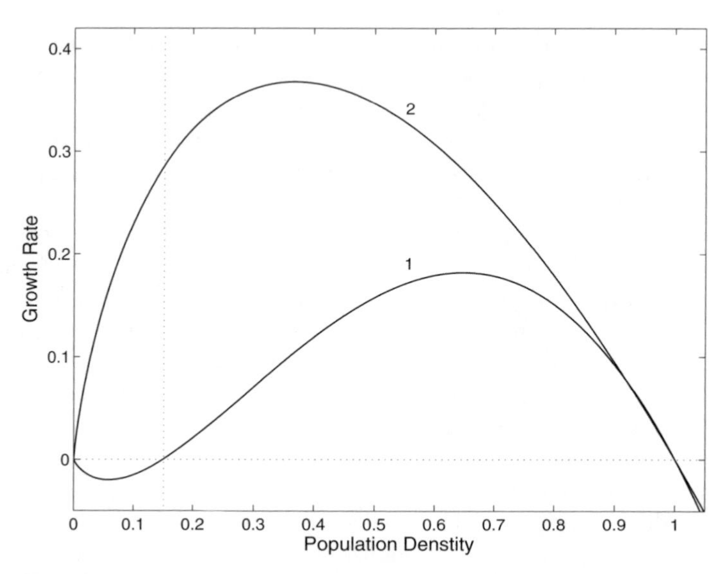

FIGURE 4.11: Different types of density-dependence in the population growth rate: curve 1 corresponds to a population affected by the strong Allee effect and curve 2 shows growth rate described by function $\bar{f}(u)$; see (4.37).

population inside the initial "patch" appears to be viable or prone to extinction depends on relation between L and L_{cr} but L is actually prescribed by the problem formulation, cf. (4.54). It seems that a more adequate approach should consider the dynamics of the introduced population in an unbounded domain and the critical radius of the invaded area should arise self-consistently from solution properties.

In terms of the single-species model (4.65), the problem of critical aggregation in an unbounded spatial domain appears most naturally in case the population local growth rate becomes negative for small values of the population density u, i.e., when $f(u) = f_c(u)$ where f_c has the following properties:

$$f_c(u) > 0 \ \text{ for } \ \epsilon < u < 1; \ \ f_c(0) = f_c(\epsilon) = f_c(1) = 0; \tag{4.67}$$

$$f_c(u) < 0 \ \text{ for } \ 0 < u < \epsilon \ \text{ and } \ u > 1, \tag{4.68}$$

e.g., see curve 1 in Fig. 4.11. This type of population dynamics may arise as a result of the strong Allee effect (see Section 1.2); in that case $\epsilon = \beta$. If, at a certain moment t_0, the maximal species density falls below the threshold density ϵ, it obviously leads to population extinction.

On the contrary, for a population with nonnegative growth rate, e.g., see (4.35), the state $u \equiv 0$ is a "repeller" and the condition $\max u(r, t_0) < \epsilon$ will never lead to the population extinction, however small the value of ϵ is. For the special case $f(u) = f_\gamma(u)$, cf. (4.37), this type of behavior is demonstrated by the exact solution (4.46), (4.47). Although at an early stage of the process the maximal species density g may fall to a small value (depending on the parameter values, see Figs. 4.6 and 4.7), in the large-time limit $g(t)$ always approaches unity (cf. Figs. 4.5 to 4.7). The radius $R(t)$ of the invaded area grows monotonically and it means that any event of local invasion (e.g., described by initial species distribution (4.39)) leads to a global invasion even for very small values of R_{in} and g_0.

However, it becomes possible to treat the problem of critical aggregation in terms of a single-species model (4.65) with nonnegative growth rate as well if we admit the "practical stability concept." According to this concept, the dynamics of a population leads to its extinction, if, at a certain time moment t_0, its maximal concentration falls below a certain small value, i.e., $\max u(r, t_0) < \epsilon$. (Note that the population is assumed to go extinct even if the rigorous solution of the problem implies that the population density will actually be growing for $t > t_0$.) Here a positive constant ϵ can be considered as a biological or environmental characteristic for a given population and plays the same role as the threshold density does for an Allee-type population.

Obviously, actual behavior of the species density is subject to the initial conditions. Speaking generally, in terms of the problem (4.65)–(4.66) it means that the issue of the population extinction/survival may depend on the details of the species initial distribution $\Phi(r)$. Mathematical consideration of the problem of critical aggregation in case of the initial condition of an arbitrary form appears to be extremely difficult. In fact, we are not aware of any rigorous result obtained for this problem. To make it treatable, we suggest that the initial distribution is not an arbitrary function of r but belongs to a certain family, so that different members of this family correspond to different values of certain parameters. Specifically, in order to make use of the exact self-similar solution obtained in Section 4.2.1, we consider the case when the initial condition is given by

$$\Phi(r) = g_0 \, \exp\left[-\left(\frac{r}{R_{in}}\right)^2\right], \tag{4.69}$$

cf. (4.39), where R_{in} and g_0 are parameters. Since the solution of (4.65) with (4.69) is known (see (4.40), (4.46) and (4.47)), the problem of predicting extinction/survival of the invading species is now reduced to the problem of finding corresponding relations between the radius and the amplitude of the initially inhabited domain.

Let us mention here that the practical stability concept, originally introduced from heuristic arguments, appears also as a result of strict mathematical consideration. Namely, it is obvious that, for sufficiently small u, any function $f_c(u)$ with the properties (4.67–4.68) can be dominated by function $\bar{f}(u) = -u \ln u$; see Fig. 4.11. More rigorously, it means that there exists a certain $U > \epsilon$ so that $f_c \le \bar{f}(u)$ for $0 \le u \le U$. Let $u_c(r, t)$ be the solution of Eq. (4.65) with the initial conditions given by (4.69) and the growth function possessing the properties (4.67–4.68). Considering the problem (4.65) with

(4.69) for $g_0 \le U$ and applying the comparison principle for the solutions of nonlinear parabolic equations (see Section 7.4), we arrive at the following relation:

$$u_c(r,t) \le u(r,t) = g(t)e^{-\xi^2} \quad \text{for any} \quad r,t \ge 0 \tag{4.70}$$

where $\xi = r/R(t)$. If it happens that $\max u(r,t_0) = g(t_0) < \epsilon$ for a certain t_0, it means that the "real" population density falls below the survival threshold as well and the population goes extinct.

Thus, in terms of exact solution (4.46–4.47) we can use the following condition of the population extinction:

$$\text{there exists such a moment } t_0 \quad \text{that} \quad g(t_0) < \epsilon \tag{4.71}$$

where ϵ is a parameter, $0 < \epsilon < 1$. Taking into account (4.47), from (4.71) we arrive at

$$\ln g_0 + e^{t_0} \ln\left(\frac{1}{\epsilon}\right) < \frac{2\sigma}{4+R_{in}^2} \ln\left[\frac{R(t_0)}{R_{in}}\right] \tag{4.72}$$

where $\sigma = 2 + 2\eta$. Therefore, the species introduction described by (4.69) leads to species extinction in case inequality (4.72) holds for a certain t_0. The critical relation $g_0 = g_{cr}(R_{in}; a, \sigma, \epsilon)$ between the problem parameters is defined by the expression (4.72) when the inequality sign is changed to the equality sign. While for a "subcritical" case $g_0 < g_{cr}$ the invasion will be unsuccessful and the invasive species goes extinct, for a "supercritical" case $g_0 > g_{cr}$ a local introduction can lead to species geographical spread.

Considering the properties of inequality (4.72), after a little algebra (details are given below) we arrive at the following critical relation:

$$g_{cr} \equiv \epsilon \quad \text{for} \quad R_{in} \ge R^* \tag{4.73}$$

and

$$g_{cr} = \exp\left(\frac{1}{4+R_{in}^2}(\zeta - 2\sigma \ln R_{in})\right) \quad \text{for} \quad R_{in} < R^* \tag{4.74}$$

where $\zeta = 2\sigma \ln R^* - \sigma + 4\ln\epsilon$ and $R^* = (\sigma/\ln(1/\epsilon))$.

Particularly, for $R_{in} \ll 1$, Eq. (4.74) takes a simpler asymptotical form:

$$g_{cr} \simeq \tilde{\zeta}\, R_{in}^{-\sigma/2} \tag{4.75}$$

where $\tilde{\zeta} = \exp(\zeta/4)$.

Thus, equations (4.73) and (4.74) give the critical relation between the maximum population density and the initially invaded domain, so that $g_0 < g_{cr}$ provides a sufficient condition for the population extinction. Fig. 4.12 shows the critical relation $g_0 = g_{cr}(R_{in})$ (logarithmic plot) obtained in 1-D

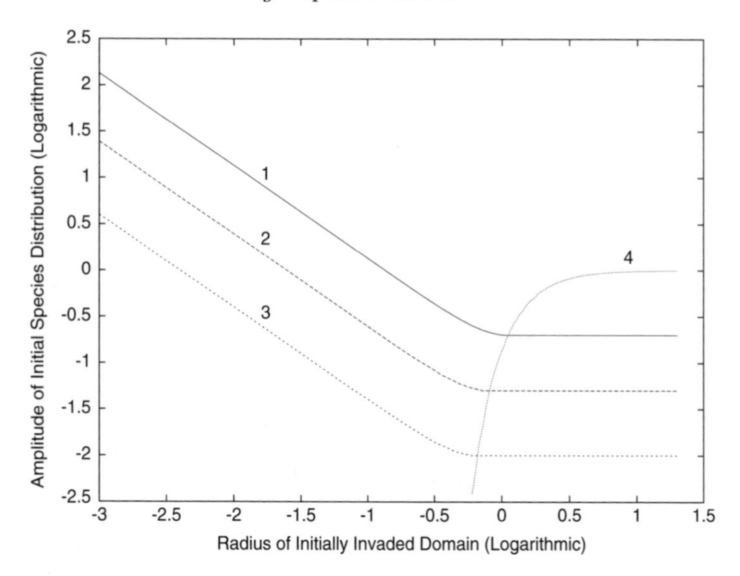

FIGURE 4.12: The critical relation between the maximum population density and the radius of initially invaded domain obtained for different values of the survival threshold ϵ, curve 1 for $\epsilon = 0.2$, curve 2 for $\epsilon = 0.05$ and curve 3 for $\epsilon = 0.01$. Parameters from the domain below each of the curves 1, 2 and 3 correspond to species extinction; parameters from the domain above each curve correspond to species invasion. Auxiliary curve 4 consists of the points where different asymptotics matches; see details in the text (with permission from Petrovskii and Shigesada, 2001).

case ($\sigma = 2$) for $\epsilon = 0.2$ (curve 1), $\epsilon = 0.05$ (curve 2) and $\epsilon = 0.01$ (curve 3). Auxiliary curve 4 consists of the points where the two branches of the critical curve meet for different values of ϵ, cf. Eqs. (4.73) and (4.74).

An important point that we want to emphasize here is that, although the critical density-radius relation (4.73–4.74) was obtained for the specific growth function $\bar{f}(u)$ and for the specific initial condition (4.69), qualitatively, the shape of the curves shown in Fig. 4.12 appears to be universal. Computer experiments accomplished for other growth functions and other initial distributions lead to very similar results, i.e., to a curve exhibiting an infinite growth for small R_{in}, approaching the horizontal line $g_0 = \epsilon$ for large R_{in}, and monotonously declining for intermediate values.

Derivation of Eqs. (4.73–4.74). Condition (4.72) does not immediately allow one to distinguish between "subcritical" and "supercritical" parameters R_{in} and g_0 because t_0 is unknown. Moreover, since Eq. (4.47) is rather complicated, it does not seem possible to obtain an explicit expression for it. Let us mention, however, that actually we do not need the value of t_0 because the only important point is whether such a moment exists or not. Considering the left-hand side and the right-hand side of (4.72) as two independent functions,

$$A(z) = \ln g_0 + e^z \ln\left(\frac{1}{\epsilon}\right) \quad \text{and} \quad B(z) = \frac{2\sigma}{4 + R_{in}^2} \ln\left[\frac{R(z)}{R_{in}}\right], \tag{4.76}$$

the existence of the moment t_0 (for given R_{in}, g_0 and ϵ) means that, for some values of the variable z, the plot of the function $B(z)$ lies higher than the plot of $A(z)$. Taking into account (4.46), it is easy to see that, for large z, $A(z) \simeq e^z$ while $B(z) \simeq z$ (constant factors are omitted), i.e., $A(z)$ increases faster than $B(z)$ for any value of parameters. Thus, the condition of the population extinction $A(z) < B(z)$ means that the plots of $A(z)$ and $B(z)$ have at least one intersection point. Furthermore, it is readily seen that, while the plot of $A(z)$ is convex, the plot of $B(z)$ is a concave curve. Different situations can be then distinguished based on the relation between $A(0)$, $B(0)$ and $A'(0)$, $B'(0)$ where the prime denotes the function's derivative with respect to its argument. Namely, if $A'(0) > B'(0)$, an intersection point exists only if $A(0) < B(0)$. Taking Eq. (4.76) into account, we arrive at the following system:

$$A(0) = \ln g_0 + \ln\left(\frac{1}{\epsilon}\right) \; < \; B(0) \; = \; 0 \; , \tag{4.77}$$

$$A'(0) = \ln\left(\frac{1}{\epsilon}\right) \; > \; B'(0) \; = \; \frac{\sigma}{R_{in}^2} \; . \tag{4.78}$$

That is, the invasive population goes extinct if $g_0 < g_{cr}$, where

$$g_{cr} \equiv \epsilon \quad \text{for} \quad R_{in} \geq R^* = \left(\frac{\sigma}{\ln(1/\epsilon)}\right)^{1/2} \; .$$

However, the curves $A(z)$ and $B(z)$ can intersect as well in the opposite case $A'(0) < B'(0)$. Accounting for (4.46) and considering how the plot of functions $A(z)$ and $B(z)$ changes with variations of parameters, it is not difficult to show that an intersection point exists when $g_0 < g_{cr}$ where the critical relation is now defined by the following equation:

$$g_{cr} = \exp\left(\frac{1}{4 + R_{in}^2}(\zeta - 2\sigma \ln R_{in})\right) \quad \text{for} \quad R_{in} < R^*$$

where $\zeta = 2\sigma \ln R^* - \sigma + 4 \ln \epsilon$.

In conclusion, let us mention that the relation $R_{in} = R^* = (-\sigma/\ln \epsilon)^{1/2}$ separating the two cases has a clear mathematical meaning. Namely, having it written as $\epsilon = \exp(-\sigma/R_{in}^2)$ and substituting it into $g_{cr} = \epsilon$, we obtain:

$$g_* = \exp\left(-\frac{\sigma}{R_{in}^2}\right); \tag{4.79}$$

see curve 4 in Fig. 4.12. Considering Eq. (4.47), it is not difficult to see that $g'(t = 0) > 0$ when $g_0 < g_*$ and $g'(t = 0) < 0$ otherwise. The form of Eqs. (4.44–4.45) evidently implies that if $g'(t=0) > 0$ then $g'(t) > 0$ for any $t > 0$. Thus, in case $g_0 < g_*$ the amplitude of the solution increases monotonically and the invasive population can only go extinct if the condition of the population extinction holds already at the beginning of the process, $g_0 < \epsilon$.

4.3.2 * The Wilhelmsson "blow-up" solution

As it has already been mentioned at the beginning of this chapter, there is growing evidence that motility of some species can be density-dependent. In Section 4.1 we made an attempt to account for this phenomenon by means of considering small-scale migrations whose intensity depends on the population density. An alternative approach is based on introducing density-dependence into the species diffusivity. A few models with nonlinear diffusion describing propagation of traveling population fronts will be considered in Chapter 5. In this section, we are going to give an insight into the possible modifications that diffusion density-dependence can bring to the problem of critical aggregation. Our analysis will be based on the exactly solvable model developed by Wilhelmsson (1988a,b).

We assume that the dynamics of the alien species is described by the following equation:

$$\frac{\partial u(x,t)}{\partial t} = \frac{\partial}{\partial x}\left(D(u)\frac{\partial u}{\partial x}\right) - Au + Bu^p - ku^q \tag{4.80}$$

where A, B, k, p and q are certain positive parameters. Apparently, when B is sufficiently large and $1 < p < q$, the reaction term in (4.80) describes the local growth of the population affected by the strong Allee effect. We also assume that density-dependence of species diffusivity is described by a power law so that $D(u) = D_0 u^\delta$.

Introducing dimensionless variables in an obvious way, from (4.80) we obtain:

$$u_t(x,t) = \left(u^\delta u_x\right)_x - u + u^p - \kappa u^q \ . \tag{4.81}$$

The initial distribution of species is $u(x,0) = \Phi(x)$ where $\Phi(x \to \pm\infty) = 0$, cf. the comments below Eq. (4.39). The form of Φ which is "intrinsic" for this problem will become clear later.

For arbitrary p, q and δ, Eq. (4.81) is unlikely to be analytically solvable. To make it integrable, some simplifications are necessary. For this purpose, we neglect the last term in the right-hand side so that (4.81) turns to

$$u_t = \left(u^\delta u_x\right)_x - u + u^p \ . \tag{4.82}$$

The properties of Eqs. (4.81) and (4.82) are essentially different for large population density because (4.82) does not possess the upper steady state

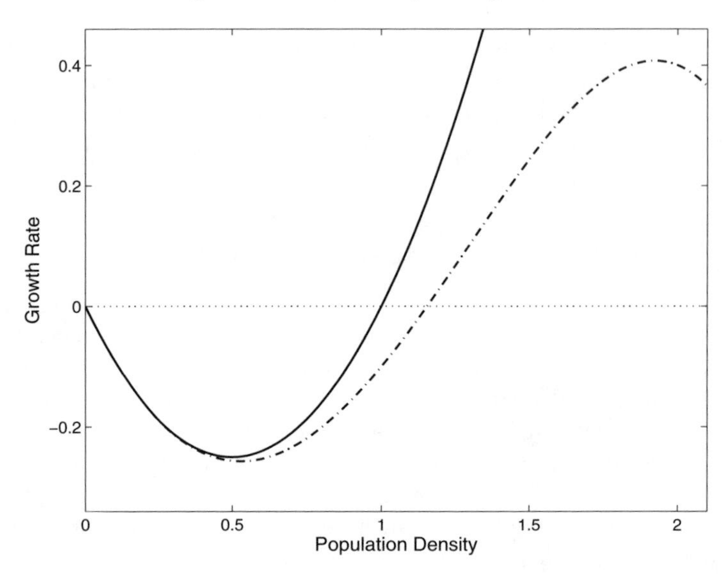

FIGURE 4.13: Full (dashed-and-dotted curve) and reduced (solid curve) parameterization of the population growth rate (see Eqs. (4.81) and (4.82), respectively.)

any more (see Fig. 4.13). However, for the values of u on the order of the threshold density the corresponding growth functions lie close to each other, cf. the dashed-and-dot and solid curves in Fig. 4.13. This is exactly what we need because, in case the radius of initially inhabited domain is not very small, the critical value of the population density is expected to be on the order of the survival threshold, e.g., see Fig. 4.12.

Moreover, it is readily seen that the growth function of (4.82) majorizes the growth function of (4.81). Thus, by virtue of the comparison principle for the solutions of nonlinear parabolic equations, cf. Section 7.4, solutions of (4.82) provide an upper bound for solutions of (4.81).

Equation (4.82) can be further simplified if we additionally assume that $p = \delta + 1$. In this case, the linear term in the right-hand side of (4.82) can be eliminated by a simple change of variables. Indeed, considering

$$u(x,t) = \tilde{U}(x,t)e^{-t}, \tag{4.83}$$

from (4.82) we first arrive at

$$e^{\delta t}\tilde{U}_t = \left(\tilde{U}^{\delta}\tilde{U}_x\right)_x + \tilde{U}^{\delta+1} \tag{4.84}$$

which immediately turns into

$$U_\tau = \left(U^{\delta}U_x\right)_x + U^{\delta+1} \tag{4.85}$$

where $U(x,\tau) = \tilde{U}(x,t(\tau))$, provided that the new variable τ satisfies the following relation:

$$e^{\delta t}\frac{d}{dt} = \frac{d}{d\tau} \ . \tag{4.86}$$

Eq. (4.86) is equivalent to

$$\frac{d\tau}{dt} = e^{-\delta t} \tag{4.87}$$

which gives

$$\tau(t) = \frac{1}{\delta}\left(1 - e^{-\delta t}\right) \tag{4.88}$$

assuming that $\tau(0) = 0$. Note that, while the original variable t is defined on $(0,\infty)$, the new variable τ is defined on a finite interval $(0, 1/\delta)$. This observation will appear to have a crucial impact on the solution properties.

Now, let us look for a solution of Eq. (4.85) in the following form:

$$U(x,\tau) = \psi(\tau)\phi(x). \tag{4.89}$$

Having substituted (4.89) into (4.85), we obtain

$$\psi^{-(\delta+1)}\,\frac{d\psi}{d\tau} \;=\; \frac{1}{\phi}\left[\frac{d}{dx}\left(\phi^{\delta}\frac{d\phi}{dx}\right) + \phi^{\delta+1}\right] \;=\; \mu. \tag{4.90}$$

As a result of the variable separations, the left-hand side of (4.90) formally depends only on τ and the right-hand side depends only on x. That actually means that each of them is equal to a certain constant which we have denoted as μ.

Thus, Eq. (4.90) is equivalent to the following system:

$$\frac{d\psi}{d\tau} \;=\; \mu\psi^{\delta+1}\ , \tag{4.91}$$

$$\frac{d}{dx}\left(\phi^{\delta}\frac{d\phi}{dx}\right) + \phi^{\delta+1} - \mu\phi \;=\; 0\ . \tag{4.92}$$

Equation Eq. (4.91) is readily solved:

$$\psi(\tau) = \left[\delta\mu(\tau_0 - \tau)\right]^{-1/\delta} \tag{4.93}$$

where τ_0 is an integration constant defined, as usual, by the initial conditions.

To find a solution of Eq. (4.92) is a more difficult problem. Let us first note that it can be written as follows:

$$\frac{d^2z}{dx^2} + pz - \mu p z^{1/p} = 0 \tag{4.94}$$

where $z = \phi^p$; for convenience, we have re-introduced the notation $\delta + 1 = p$. Considering $dz/dx = w(z)$, Eq. (4.94) takes the form

$$w\frac{dw}{dz} = \mu p z^{1/p} - pz \tag{4.95}$$

which is easily integrated yielding

$$w(z) \quad = \quad \frac{dz}{dx} \quad = \quad \left(\frac{2\mu p^2}{p+1} z^{1+1/p} - pz^2 + C\right)^{1/2} . \tag{4.96}$$

Introduction of an exotic species normally takes place locally; that means that $\phi(x \to \pm\infty) = 0$. Correspondingly, $dz/dx = 0$ for $x \to \pm\infty$. Then, substituting $w(0) = 0$ into Eq. (4.96) we obtain that $C = 0$.

Evidently, in order to obtain the solution of Eq. (4.96) we must calculate the following integral:

$$\int \left[z^{1+1/p}\left(\frac{2\mu p^2}{p+1} - pz^{1-1/p}\right)\right]^{-1/2} dz \ . \tag{4.97}$$

Although at first sight it may seem rather scary, by means of the substitution

$$\left(\frac{2\mu p^2}{p+1} - pz^{1-1/p}\right) = y^2 \ , \tag{4.98}$$

integral (4.97) is reduced to a standard one:

$$-\frac{2\sqrt{p}}{p-1}\int \left(\frac{2\mu p^2}{p+1} - y^2\right)^{-1/2} dz = \\ -\frac{2\sqrt{p}}{p-1}\arcsin\left(y\sqrt{\frac{p+1}{2\mu p^2}}\right) . \tag{4.99}$$

Thus, from (4.96–4.99) we obtain

$$\frac{2\sqrt{p}}{p-1}\arcsin\left(y\sqrt{\frac{p+1}{2\mu p^2}}\right) = C_1 - x \tag{4.100}$$

where C_1 is another integration constant. Taking into account definition of variables y and z, (4.100) turns to

$$\phi^{p-1}(x) = \frac{2\mu p}{p+1}\cos^2\left[\frac{(p-1)}{2\sqrt{p}}(C_1 - x)\right] . \tag{4.101}$$

Since the original equation does not contain x explicitly, its solution appears to be invariant with respect to a shift along axis x. Therefore, constant C_1 defines where the maxima of $\phi(x)$ are situated. In case we assume, for convenience, that one of them is reached at $x = 0$, then $C_1 = 0$ and we finally arrive at the solution of Eq. (4.92):

$$\phi(x) = \left(\frac{2\mu p}{p+1}\right)^{1/(p-1)} \cos^{2/(p-1)}\left[\frac{(p-1)}{2\sqrt{p}}x\right]. \tag{4.102}$$

Coming back to original variables, from (4.83), (4.89), (4.93) and (4.102), and also recalling that $p = \delta + 1$, we obtain the solution of Eq. (4.82):

$$u(x,t) = \left[\frac{2(\delta+1)e^{-\delta t}}{\delta(\delta+2)(\tau_0 - \tau)}\right]^{1/\delta} \cos^{2/\delta}\left(\frac{\delta x}{2\sqrt{\delta+1}}\right) \tag{4.103}$$

where $\tau(t)$ is given by (4.88).

Let us note that the function $u(x,t)$ given by (4.103) is periodical in space with period $\Lambda = (2\pi/\delta)\sqrt{\delta+1}$ at any moment t, in particular, for $t = 0$. Since biological invasion starts with a local event, a periodical function can hardly be used to describe species introduction. However, we can construct a single-hump solution by means of the following relations:

$$u(x,t) = A(t)\cos^{2/\delta}\left(\frac{\pi x}{\Lambda}\right) \quad \text{for} \quad |x| < \frac{\Lambda}{2}\,, \tag{4.104}$$

$$u(x,t) \equiv 0 \quad \text{for} \quad |x| \geq \frac{\Lambda}{2} \tag{4.105}$$

where

$$A(t) = \left[\frac{2(\delta+1)e^{-\delta t}}{\delta(\delta+2)(\tau_0 - \tau)}\right]^{1/\delta} \tag{4.106}$$

is the maximum population density.

It is readily seen that the solution (4.104–4.106) is a "blow-up" type of solution because it turns to infinity at finite time $\bar{t}$ when $\tau(\bar{t}) = \tau_0$. (More about the blow-up solutions and their properties can be found in Samarskii et al. (1987) and Bebernes and Eberly (1989).) Now, we recall that $\tau \in (0, 1/\delta)$ and whether the solution actually has the singularity or not depends on the value of τ_0. In case $\tau_0 > 1/\delta$, the solution will never blow-up. In its turn, τ_0 depends on the maximum population density in the initial spatial distribution of the introduced species. Indeed, letting $t = 0$ from (4.106) we obtain:

$$\tau_0 = \frac{2(\delta+1)}{\delta(\delta+2)}A_0^{-\delta} \tag{4.107}$$

where $A_0 = A(0)$. From the critical relation $\tau_0 = 1/\delta$, we then immediately arrive at the equation for the critical population density:

$$A_{cr} = \left(\frac{2(\delta+1)}{\delta+2}\right)^{1/\delta} . \tag{4.108}$$

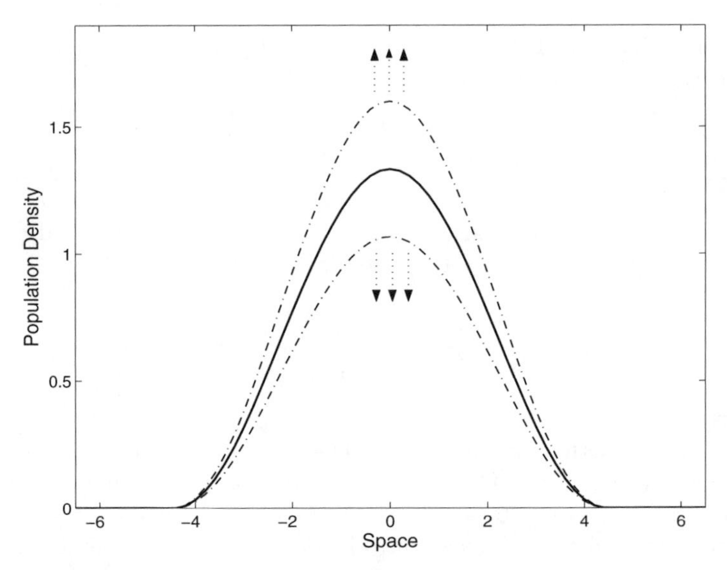

FIGURE 4.14: The critical initial profile of the population density (solid curve) as given by (4.109–4.110). The profiles that lie above and below the critical one, cf. the dashed-and-dotted curves, correspond to species invasion and species extinction, respectively.

Thus, the initial species distribution $u(x,0) = \hat{\Phi}(x)$ where

$$\hat{\Phi}(x) = A_{cr} \cos^{2/\delta}\left(\frac{\pi x}{\Lambda}\right) \quad \text{for} \quad |x| < \frac{\Lambda}{2}\,, \tag{4.109}$$

$$\hat{\Phi}(x) \equiv 0 \quad \text{for} \quad |x| \geq \frac{\Lambda}{2} \tag{4.110}$$

is a separatrix separating the solutions tending to zero (for $A_0 < A_{cr}$) from the solutions tending to infinity (for $A_0 > A_{cr}$); see Fig. 4.14. Apparently, the decaying solutions correspond to species extinction. As for the singular solutions, their growth can be interpreted as the beginning of species spread, especially if we take into account that their unboundedness is clearly a consequence of our neglecting the term u^q in the original equation (4.81). It should be also mentioned that, since solutions of (4.82) majorize solutions of (4.81), from the standpoint of the full Eq. (4.81), the condition $\Phi(x) \leq \hat{\Phi}(x)$ is a sufficient condition of alien species extinction.

In conclusion, we want to mention that, as well as in the Kierstead–Slobodkin model, solution (4.104–4.106) describes the population dwelling in a finite domain. However, an important distinction is that in this case the length of the domain appears as an intrinsic property of the solution, not being fixed by the problem formulation. Formally, Eq. (4.82) is considered in an unbounded space and the dynamical "confinement" of the alien population is a result of the diffusion density-dependence.

Chapter 5

Density-dependent diffusion

In the previous chapters, our analysis was mostly focused on the cases when species diffusivity is constant. There is, however, growing evidence that in some ecological situations species motility can depend on the population density (Gurney and Nisbet, 1975; Hengeveld, 1989). There are different manifestations of this phenomenon and different modeling approaches to take it into account (Turchin, 1998; Ardity et al., 2001). Here we assume that density-dependence of species motility can be adequately described by means of variable diffusivity. Our goal is to investigate what changes density-dependent diffusion can bring into the system dynamics.

In a single-species model, the effects of diffusivity dependence on the population density can be taken into account by means of an immediate extension of the basic equation, i.e.:

$$u_t(x,t) = (D(u)u_x)_x + F(u) \ , \tag{5.1}$$

cf. Section 2.1. Apparently, the solution properties depend on the form of $D(u)$. Inferences from ecological observations indicate that, in those cases when the density-dependence takes place, $D(u)$ is likely to be an increasing function. In this chapter, we are going to give an insight into consequences of diffusivity density-dependence by considering a few exactly solvable models with different $D(u)$ and $F(u)$. Our consideration will be mostly based on original works by Herrera et al. (1992) and Strier et al. (1996).

5.1 The Aronson–Newman solution and its generalization

The following equation has been introduced by Gurney and Nisbet (1975, 1976) as the simplest model of a population with density-dependence motility:

$$u_t(x,t) = (uu_x)_x + u(1-u) \tag{5.2}$$

(in dimensionless variables). Invasion of an alien species corresponds to the special choice of conditions at infinity:

$$u(x \to -\infty, t) \ = \ 0, \qquad u(x \to \infty, t) \ = \ 1 \tag{5.3}$$

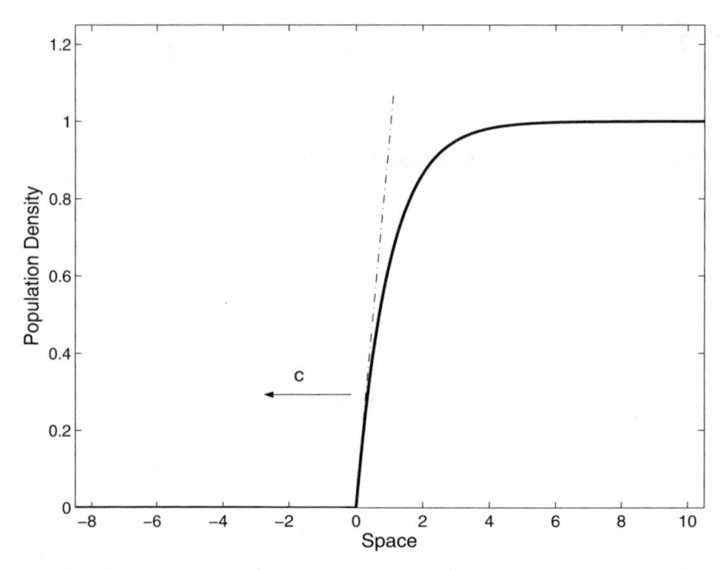

FIGURE 5.1: A "sharp" traveling population front described by exact solution (5.4–5.5). Note that the population density gradient has a discontinuity at the leading edge.

where we assume without any loss of generality that the invasive species spreads from right to left.

Aronson (1980) and Newman (1980) considered independently the properties of Eq. (5.2) and showed that it has an exact solution describing a traveling population front:

$$U(\xi) = 1 - \exp\left(-\frac{\xi - \xi_0}{\sqrt{2}}\right) \quad \text{for} \quad \xi \geq \xi_0 \,, \tag{5.4}$$

$$U(\xi) = 0 \quad \text{for} \quad \xi < \xi_0 \tag{5.5}$$

where $\xi = x - ct$ and c is the speed of the wave. The profile described by (5.4–5.5) is shown in Fig. 5.1. The remarkable feature of this solution is that the population density is greater than zero only in a semi-infinite interval, i.e., for $\xi \geq \xi_0$.

Note that the "leading edge" $\xi = \xi_0$ is a singular point where the solution first derivative has a discontinuity; for that reason, the wave described by (5.4–5.5) is often referred to as a "sharp" front. Correspondingly, the derivative $dU/d\xi$ does not exist at the leading edge and thus, strictly speaking, Eq. (5.2) may not be applicable when $\xi = \xi_0$. In a more rigorous sense, $U(\xi)$ given by (5.4–5.5) should be regarded as a "weak" solution, i.e., a solution of the corresponding integral-differential equation (Herrera et al., 1992; for a more general discussion see Volpert and Khudyaev, 1985). However, it still appears possible to treat (5.4–5.5) in terms of Eq. (5.2) if we take into account that

$U(\xi)$ and the flux $D(U)dU/d\xi$ have no discontinuity at $\xi = \xi_0$.

There are different ways to arrive at (5.4–5.5). Apparently, traveling fronts arise as solutions to the following equation:

$$\frac{d}{d\xi}\left(U\frac{dU}{d\xi}\right) + c\frac{dU}{d\xi} + U(1-U) = 0 \ . \tag{5.6}$$

Since Eq. (5.6) does not contain variable ξ explicitly, its order can be reduced by means of introducing a new variable $dU/d\xi = \psi(U)$, cf. Section 9.1. Equation (5.6) then takes the form

$$\psi\frac{d}{dU}(U\psi) + c\psi + U(1-U) = 0 \ . \tag{5.7}$$

Aronson (1980) and Newman (1980) (see also (Murray, 1989) for a brief review of their results) studied the phase plane (U, ψ) of Eq. (5.6) and found that, for a certain value of c, there must exist a unique trajectory connecting the equilibrium points $(1, 0)$ and $(0, -c)$. Assuming that the trajectory is given by a straight line, $\psi = -c(1-U)$, they then were able to find the speed of the wave c and, eventually, to obtain the solution (5.4–5.5).

Here we use another way to arrive at (5.4–5.5) which is more consistent with our previous analysis. Let us try to look for a solution of (5.7) in the form of a polynomial of m-th order. To obtain possible values of m, we first consider $\psi \sim U^m$. Having substituted it into the equation, we obtain:

$$U^{2m} + U^{2m} + U^m + U - U^2 = 0 \tag{5.8}$$

where we have omitted the coefficients. Evidently, different powers of U match each other only if $m = 1$. A general form of the first-order polynomial would be $\psi = \alpha - AU$ (where α and A are certain coefficients) which we regard as an ansatz. We then take into account that $U = 1, \psi = 0$ is an equilibrium point (corresponding to the condition at $\xi \to \infty$) so that $\psi(1) = 0$ and thus $A = \alpha$. Substituting $\psi = \alpha(1-U)$ into (5.7), we obtain:

$$-\alpha^2 U + \alpha^2(1-U) + cU + U = 0 \ . \tag{5.9}$$

Matching different powers of U, from (5.9) we arrive at

$$c = -\alpha \ , \qquad \alpha = \pm\frac{1}{\sqrt{2}} \ . \tag{5.10}$$

Here the choice of the sign for α must be consistent with the choice of conditions at infinity. In the case of (5.3), $U(\xi)$ is an increasing function so that $\psi = dU/d\xi > 0$ and $\alpha > 0$. Thus, $\alpha = 1/\sqrt{2}$. Correspondingly, $c = -1/\sqrt{2} < 0$ so that the front travels to the left.

To obtain the shape of the wave profile, we make use of the ansatz itself:

$$\psi = \frac{dU}{d\xi} = \frac{1-U}{\sqrt{2}} \ . \tag{5.11}$$

From (5.11), we immediately obtain:

$$U(\xi) = 1 - \exp\left(-\frac{\xi - \xi_0}{\sqrt{2}}\right). \tag{5.12}$$

Note that solution (5.12) by itself does not possess any clear biological meaning because it is not nonnegative if considered on the whole line. Strictly speaking, it makes sense only for $\xi \geq \xi_0$ where $U \geq 0$. However, we recall that $U(\xi) \equiv 0$ is also a solution. Thus, a biologically meaningful solution can be obtained by inosculating the two different branches. As it was mentioned above, this corresponds to a weak solution; alternatively, we can consider Eq. (5.6) in two semi-infinite domains and match different solutions at the interface $\xi = \xi_0$ by imposing the conditions of continuity on the population density and its flux. From here, we arrive at (5.4–5.5).

A point of interest is the biological relevance of the Aranson–Newman solution. One of the features of the diffusion-reaction equations with constant diffusivity that they have often been criticized for is that they, actually, describe an infinite-speed propagation of the leading edge. Indeed, for initial conditions of compact support at $t = 0$, they predict the population density to be positive in the whole space for any $t > 0$, however small t is. This is readily seen from the general solution of the diffusion equation (see Section 9.3) and is extended to the diffusion-reaction equation with a logistic growth by virtue of the comparison principle. Although at large distances the density is exponentially small, theoretically speaking, the probability of catching an individual of the invasive species appears to be greater than zero at any position in space which is not realistic. On the contrary, Eq. (5.2) predicts that, in front of the propagating population front, the invasive species is absent in the strict mathematical sense. It remains unclear, however, whether this feature is a specific property of Eq. (5.2) with the diffusion coefficient being proportional to the population density or a more general property of density-dependent diffusion. This question is addressed below.

5.1.1 A general case

In this section, we consider how the Aronson–Newman solution can be generalized to take into account other cases of density-dependence. Specifically, we consider the following equation:

$$u_t(x,t) = \left(u^\delta u_x\right)_x + u^p - u^k\ . \tag{5.13}$$

For biological reasons, parameters p, k are positive, δ is nonnegative and $k > p$ in order to ensure stability of the upper steady state $u = 1$.

As above, we are interested in traveling wave solutions, i.e., solutions of the equation

$$\frac{d}{d\xi}\left(U^\delta \frac{dU}{d\xi}\right) + c\frac{dU}{d\xi} + U^p - U^k = 0 \tag{5.14}$$

where $U = U(\xi)$, $\xi = x - ct$ and c is the speed of the wave. Since (5.14) does not contain ξ explicitly, it is convenient to introduce a new variable $dU/d\xi = \psi(U)$. Eq. (5.14) is then reduced to the following equation of the first order:

$$\psi\left(U^{\delta}\frac{d\psi}{dU} + \delta U^{\delta-1}\psi + c\right) + U^{p}\left(1 - U^{k-p}\right) = 0\ . \tag{5.15}$$

We will try to obtain exact solutions of (5.15) by means of introducing a relevant ansatz. Note that our goal here is to obtain a generalization of the Aranson–Newman solution to the case given by Eq. (5.14). Correspondingly, we expect it to describe a sharp front, i.e., a profile of population density that turns to zero with a finite derivative. Thus, a relevant choice of ansatz seems to be

$$\psi(U) = \alpha\left(1 - U^{m}\right) \tag{5.16}$$

(where α and m are to be defined) so that $\psi(1) = 0$ and $\psi(0) = \alpha \neq 0$ gives the gradient of the wave profile at the leading edge; see Fig. 5.1.

Having substituted (5.16) into (5.15), after some transformations we obtain:

$$\alpha\left(1 - U^{m}\right)\left[c + \alpha\delta U^{\delta-1} - \alpha(m+\delta)U^{m+\delta-1}\right] \tag{5.17}$$
$$+ U^{p}\left(1 - U^{k-p}\right) = 0\ .$$

The choice of the ansatz is appropriate if, after substitution of (5.16) into (5.15), we are able to find all the coefficients by means of equating different powers of U. However, since we actually have only two unknown values to be determined in this way, i.e., α and c, the number of corresponding algebraic equations must not be larger than two. In its turn, it means that the equation can contain U only in two different powers. Evidently, for arbitrary p, k, δ and m, Eq. (5.17) contains more than two powers. Therefore, we have to make certain suggestions about parameter values in order to simplify the equation. In particular, we notice that, if $k - p = m$, Eq. (5.17) takes a simpler form:

$$\alpha c + \alpha^{2}\delta U^{\delta-1} - \alpha^{2}(m+\delta)U^{m+\delta-1} + U^{p} = 0\ . \tag{5.18}$$

For arbitrary p, δ and m, Eq. (5.18) still contains four different powers, i.e., 0, $\delta - 1$, $m + \delta - 1$ and p. Here, the term (αc) can only be matched with another one if either (**i**) $\delta - 1 = 0$ or (**ii**) $\delta + m - 1 = 0$.

Let us first consider (**ii**). In this case, Eq. (5.18) takes the form

$$\alpha[c - \alpha(m+\delta)] + \alpha^{2}\delta U^{\delta-1} + U^{p} = 0\ . \tag{5.19}$$

It is readily seen that the only remaining step to be made is to let $p = \delta - 1$ in order to match the last two terms in the left-hand side of (5.19). However, equating the coefficients, we then obtain $\alpha^{2}\delta + 1 = 0$ which is impossible because both α and δ are nonnegative. Thus, case (**ii**) does not lead to a solution.

In contrast, in the case $\delta = 1$ we immediately obtain that $m = p$ and $k = 2m$. Matching different powers in Eq. (5.18), we arrive at the equation for the speed of the wave:

$$c = -\alpha , \quad \alpha = \pm \frac{1}{\sqrt{p+1}} \tag{5.20}$$

where different signs for α correspond to different conditions at infinity, i.e., whether the invasive species spreads along axis x or against axis x. For the conditions given by (5.3), $U(\xi)$ must be an increasing function; therefore, $\alpha > 0$.

To obtain the wave profile, we have Eq. (5.16) which now reads as follows:

$$\frac{dU}{d\xi} = \frac{1}{\sqrt{p+1}} (1 - U^p) . \tag{5.21}$$

For an arbitrary p, Eq. (5.21) does not lead to an explicit expression for $U(\xi)$. However, at least two cases when the solution can be obtained in a closed form are easily identified. One of them is given by $p = 1$ ($k = 2$) when (5.21) leads to the Aranson–Newman solution (5.4).

The second integrable case is given by $p = 2$ ($k = 4$). Correspondingly, from (5.21) we obtain:

$$U(\xi) = \tanh\left(\frac{\xi - \xi_0}{\sqrt{3}}\right) . \tag{5.22}$$

The right-hand side of (5.22) is nonnegative only for $\xi \geq \xi_0$; therefore, we have to inosculate it with the trivial solution $U(\xi) \equiv 0$ in the same manner as it was done for the Aranson–Newman solution:

$$U(\xi) = \tanh\left(\frac{\xi - \xi_0}{\sqrt{3}}\right) \quad \text{for} \quad \xi \geq \xi_0 , \tag{5.23}$$

$$U(\xi) = 0 \quad \text{for} \quad \xi < \xi_0 . \tag{5.24}$$

Compared to the Aranson–Newman solution, exact solution (5.23–5.24) covers a different case of density-dependence in the population growth when the growth function $F(u)$ is not convex. In biological terms, it corresponds to a particular case of the weak Allee effect. As well as (5.4–5.5), the solution (5.23–5.24) describes a sharp population front with the population density being greater than zero in a semi-infinite domain behind the leading edge $\xi = \xi_0$. Let us mention, however, that (5.23–5.24) has been obtained for diffusivity density-dependence the same as in the original Eq. (5.6); thus, the question of generality has yet remained open.

Remarkably, the above two cases are not the only ones when Eq. (5.14) has an exact solution. In order to comprise other cases, we need to modify the form of ansatz. The ansatz (5.16) apparently implies that the slope of the profile at the leading edge is finite, $\psi(U=0)=\alpha$. Theoretically speaking, however, we cannot rule out existence of the sharp fronts with the profile approaching zero with infinite slope, e.g., see Fig. 5.2a. To take this type of wave into account, we consider the following generalization of (5.16):

$$\psi(U)=\alpha\left(\frac{1}{U^{\gamma}}-U^{m}\right) \tag{5.25}$$

where γ is another parameter to be determined. In case $\gamma=0$, ansatz (5.25) coincides with (5.16). In case $\gamma<0$, $\psi(0)=0$ which is a typical wave profile behavior for density-independent diffusion. However, for $\gamma>0$, it possesses the required property, i.e., $\psi=dU/d\xi\to\infty$ when $U\to 0$.

Having substituted (5.25) into Eq. (5.15), we obtain:

$$\begin{aligned}\alpha U^{-\gamma}\left(1-U^{m+\gamma}\right)\left[c+\alpha\delta U^{\delta-1}\left(U^{-\gamma}-U^{m}\right)-\alpha U^{\delta}\left(\gamma U^{-\gamma-1}+mU^{m-1}\right)\right]&\\ +U^{p}\left(1-U^{k-p}\right)\ =\ 0\ .&\end{aligned} \tag{5.26}$$

As above, in order to keep the system not overdetermined, we should somehow decrease the number of different powers of U in Eq. (5.26). For that purpose, we assume that $m+\gamma=k-p$; Eq. (5.26) is then reduced to

$$\alpha\left[c+\alpha(\delta-\gamma)U^{\delta-\gamma-1}-\alpha(\delta+m)U^{\delta+m-1}\right]+U^{p+\gamma}=0\ . \tag{5.27}$$

A nontrivial solution is only possible when either (**iii**) $\delta-\gamma-1=0$ or (**iv**) $\delta+m-1=0$. In the latter case, by means of matching different powers of U, from (5.27) we arrive at the following system:

$$\delta+m-1\ =\ 0, \tag{5.28}$$

$$\delta-\gamma-1\ =\ p+\gamma, \tag{5.29}$$

$$m+\gamma\ =\ k-p \tag{5.30}$$

which leads to $p+m+2\gamma=0$. However, $p>0$ and m and γ are nonnegative; therefore, the system (5.28–5.30) is inconsistent.

In case (**iii**), instead of (5.28–5.30) we obtain the system

$$\begin{aligned}\delta-\gamma-1\ &=\ 0,\\ \delta+m-1\ &=\ p+\gamma,\\ m+\gamma\ &=\ k-p\end{aligned}$$

which is immediately solved:

$$\gamma\ =\ \delta-1,\qquad m\ =\ p,\qquad k\ =\ 2p+\delta-1. \tag{5.31}$$

Correspondingly, for the coefficients we obtain

$$\alpha c+\alpha^{2}(\delta-\gamma)=0\ , \tag{5.32}$$

$$-\alpha^{2}(\delta+m)+1=0 \tag{5.33}$$

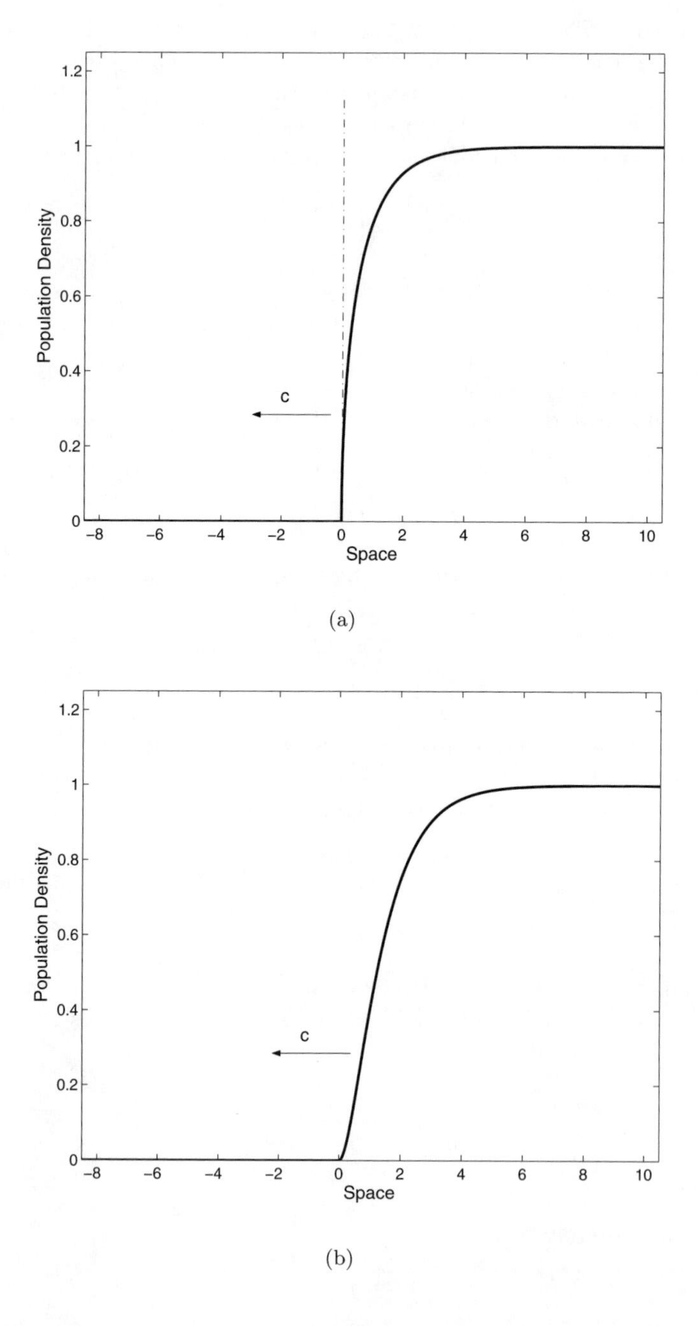

FIGURE 5.2: (a) A "sharp" traveling front with infinite steepness at the leading edge, cf. exact solution (5.36–5.37) for $\delta > 1$; (b) a traveling front with a smooth profile, cf. (5.36–5.37) for $\delta < 1$.

so that the speed of the wave is given as

$$c = -\alpha , \quad \alpha = \frac{1}{\sqrt{p+\delta}} \tag{5.34}$$

where we have taken into account the conditions at infinity; see (5.3).

The wave profile is obtained from Eq. (5.25) which now reads as follows:

$$\frac{U^{\delta-1}dU}{1-U^{p+\delta-1}} = \frac{d\xi}{\sqrt{p+\delta}} \ . \tag{5.35}$$

For arbitrary p and δ, Eq. (5.35) does not lead to a solution in a closed form. However, at least in two cases an explicit exact solution can be found easily.

First, consider $p = 1$. By virtue of (5.31), $k = \delta + 1$; thus, Eq. (5.14) describes a population with a generalized logistic growth. Eq. (5.35) is then readily solved leading to a generalization of the Aranson–Newman solution:

$$U(\xi) = \left[1 - \exp\left(-\frac{\xi - \xi_0}{\sqrt{\delta+1}}\right)\right]^{1/\delta} \quad \text{for} \ \ \xi \geq \xi_0 \ , \tag{5.36}$$

$$U(\xi) = 0 \ \ \text{for} \ \ \xi < \xi_0 \tag{5.37}$$

where we have taken into account that $U(\xi)$ cannot be negative.

Depending on the value of δ, solution (5.36–5.37) has somewhat different properties at the leading edge, i.e., in a vicinity of $\xi = \xi_0$. Namely, for $\delta = 1$, it describes a sharp front with a finite gradient, cf. Fig. 5.1. For $\delta > 1$, it describes a sharp front with an infinite gradient (see Fig. 5.2a). However, for $\delta < 1$ the gradient of the population density at the leading edge is zero (Fig. 5.2b). The latter case thus formally coincides with the case of density-independent diffusion when $dU/d\xi \to 0$ for $U \to 0$, cf. Chapter 3. Interestingly, in spite of this similarity, solution (5.36–5.37) is positively defined not on the whole line but only in a semi-finite domain.

The second integrable case is obtained for $p = \delta + 1$ ($k = 3\delta + 1$). Since $\delta > 0$, the growth function $F(u)$ is not convex; thus, it corresponds to a population with a weak Allee effect. In this case, Eq. (5.35) has the following solution:

$$U(\xi) = \left[\tanh\left(\frac{\xi - \xi_0}{\sqrt{2\delta+1}}\right)\right]^{1/\delta} \quad \text{for} \ \ \xi \geq \xi_0 \ , \tag{5.38}$$

$$U(\xi) = 0 \ \ \text{for} \ \ \xi < \xi_0 \ . \tag{5.39}$$

Clearly, the solution properties at the leading edge for different δ are the same as for (5.36–5.37).

Thus, we have shown that semi-finiteness of traveling wave solutions is a typical property of diffusion-reaction systems with density-dependent diffusivity provided $D(u) \sim u^{\delta}$. On the contrary, the "sharpness" of the traveling fronts is not an immanent property of nonlinear diffusion and is only observed for $\delta \geq 1$. These conclusions are made based on the properties of exact traveling wave solutions.

In conclusion to this section, we want to mention that, although we were able to obtain exact solutions in a closed form only for a few special cases, the equation for the speed of the traveling front (5.34) is available even when solution is unknown. A question important for applications is what speed of the front is actually observed. Although an analytical investigation of this problem is largely absent [but see Sherratt and Marchant (1996) for an example of asymptotical analysis], numerical experiments show that initial conditions of compact support always relax to the front propagating with the speed given by (5.34), cf. Herrera et al. (1992).

5.2 Stratified diffusion and the Allee effect

In the previous section we showed that introduction of density-dependence into species diffusivity can essentially modify the pattern of species spread. Interestingly, although the diffusivity was assumed to vanish when the population density tends to zero, the models considered above do not predict a possibility of invasion blocking: the speed of the propagating population front does not change its sign, cf. (5.34). The question that remains open is whether this conclusion is a consequence of the particular parameterization of diffusivity density-dependence as a power law, i.e., $D(u) \sim u^\delta$, or it is a more general result.

It appears that the condition of wave blocking for an invasive population with density-dependent motility can be obtained for a rather general case. Apparently, of primary interest are the traveling wave solutions of Eq. (5.1), i.e., the solutions of the following equation:

$$\frac{d}{d\xi}\left(D(U)\frac{dU}{d\xi}\right) + c\frac{dU}{d\xi} + F(U) = 0 \ . \tag{5.40}$$

Assuming that the invading species spreads from left to right, we set the following conditions at infinity:

$$U(\xi) = K \text{ for } \xi \to -\infty \ , \qquad U(\xi) = 0 \text{ for } \xi \to +\infty \tag{5.41}$$

where K is the population carrying capacity. Invasion corresponds to traveling fronts with $c > 0$.

For the single-species model with $D = const$, the condition of wave blocking was considered in Section 2.1. Eq. (2.22) can be readily generalized to the case of density dependent diffusivity. Indeed, multiplying Eq. (5.40) by

$D(U)dU/d\xi$ and integrating over space, we arrive at

$$\frac{1}{2}\left[\left(D(U)\frac{dU}{d\xi}\right)^2\right]_{-\infty}^{\infty} + c\int_{-\infty}^{\infty} D(U)\left(\frac{dU}{d\xi}\right)^2 d\xi - \int_0^K D(U)F(U)dU = 0\ . \tag{5.42}$$

By virtue of the conditions at infinity, $dU/d\xi = 0$ for $\xi \to \pm\infty$; therefore, the first term on the left-hand side is equal to zero. Then, from (5.42) we obtain:

$$c = \int_0^K D(U)F(U)dU \cdot \left[\int_{-\infty}^{\infty} D(U)\left(\frac{dU}{d\xi}\right)^2 d\xi\right]^{-1} . \tag{5.43}$$

Equation (5.43) cannot be used to calculate the speed in a general case because the solution $U(\xi)$ is unknown. However, it does give important information regarding the sign of the wave speed. Since the integral in the square brackets is positive, $c = 0$ means that

$$\int_0^K D(U)F(U)dU = 0\ . \tag{5.44}$$

Obviously, this is possible only if $F(U)$ changes its sign in the interval $(0, K)$. From a biological standpoint, the most interesting case of alternating-sign $F(U)$ corresponds to the strong Allee effect when the growth rate becomes negative for small values of population density; see (1.13–1.14). Thus, invasion can be blocked ($c = 0$) or turned to retreat ($c < 0$) only if the population growth of the invasive species is damped by the strong Allee effect. Moreover, Eqs. (5.43–5.44) clearly predict that diffusivity density-dependence can break the "standard" condition of wave blocking, $\int_0^K F(U)dU = 0$, cf. (2.22). It can be expected that, depending on whether $D(U)$ is an increasing or decreasing function, diffusion density-dependence either enhances species invasion or enhances species retreat, respectively. In order to address these issues in a more quantitative way, below we will consider an exactly solvable model that takes into account the Allee effect along with diffusivity density-dependence.

Note that the generalized Aronson–Newman model, cf. (5.13), gives a special case of the density-dependence when the diffusivity tends to zero for small population density. In a real ecological situation, it is more likely to happen that it tends not to zero but to a certain relatively small value D_0 (see Fig. 5.3). An appropriate parameterization could be as follows:

$$D = D_0 + AU^\gamma \quad \text{or} \quad D = D_0 + \frac{U^\gamma}{U^\gamma + A^\gamma}(D_1 - D_0) \tag{5.45}$$

where A and γ would be certain positive parameters. From the ecological standpoint, these types of density-dependence correspond to the "stratified

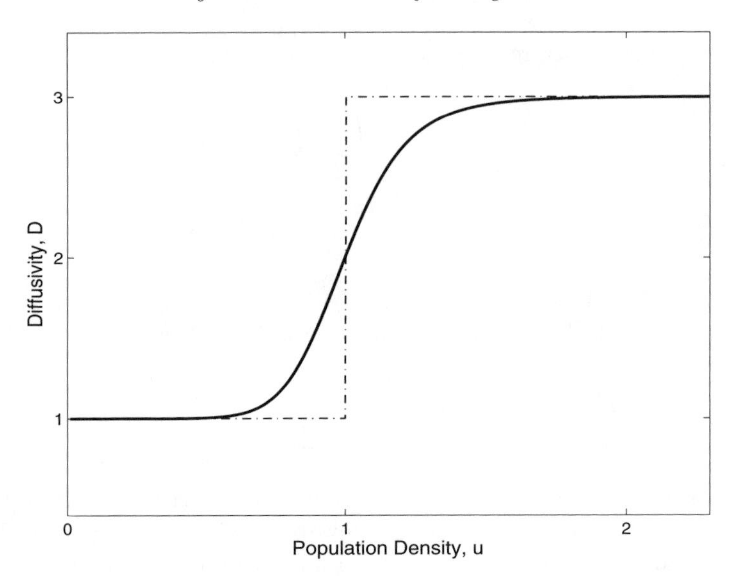

FIGURE 5.3: A sketch of diffusivity dependence on population density (solid curve) in the case of "stratified diffusion." The transition zone between the low-diffusion and high-diffusion regions is situated around $U_D = 1$; the dashed-and-dotted line shows corresponding piecewise linear approximation.

diffusion" (Hengeveld, 1989; Shigesada and Kawasaki, 1997) when species diffusivity increases along with its population density. From these two parameterizations, the second one looks biologically more reasonable taking into account that species diffusivity is related to individual mobility which cannot increase unboundedly.

However, we are not aware of any exactly solvable model where diffusivity dependence on U is of the type (5.45) or qualitatively similar. Instead, another approach can be applied. We assume that there are two ranges of the density-dependence, namely, the range of small U where the characteristic value of diffusivity is D_0, and the range of large U where the characteristic value is D_1. We further assume that the transition region between these two ranges is relatively narrow and situated around a certain population density U_D (see Fig. 5.3). The diffusivity density-dependence can then be parameterized as follows:

$$D(U) = D_0 \text{ for } U < U_D \ , \quad D(U) = D_1 \text{ for } U \geq U_D \tag{5.46}$$

or

$$D(U) = D_0 + (D_1 - D_0)\theta\,(U - U_D) \tag{5.47}$$

where

$$\theta(z) \ = \ 0 \text{ for } z < 0 \quad \text{and} \quad \theta(z) \ = \ 1 \text{ for } z \geq 0 \ . \tag{5.48}$$

The above assumptions are still not sufficient to make equation (5.40) exactly solvable. One way to make it integrable is by means of making additional assumptions about the type of density-dependence in the population growth rate. In particular, similarly to the above arguments, we assume that the transition region between the range of small U where the Allee effect strongly manifests itself and the range of large U where its impact is likely to be negligible is relatively narrow and situated at a certain density U_A. Then the growth rate can be parameterized as follows:

$$F(U) = -\alpha U + \delta\theta(U - U_A) \tag{5.49}$$

(cf. Section 3.3.2) where α and δ are positive parameters. Correspondingly, $F < 0$ for $0 < U < U_A$ and for $U > \delta/\alpha$ and $F < 0$ for $U_A < U < \delta/\alpha$, δ/α being the carrying capacity.

Introducing dimensionless variables

$$t' = t\alpha, \quad x' = x(\alpha/D_0)^{1/2}, \quad U' = U\alpha/\delta \tag{5.50}$$

and omitting primes for convenience, we arrive at equation (5.40) where now

$$F(U) = -U + \theta(U - \beta), \tag{5.51}$$

$$D(U) = 1 + (\epsilon - 1)\theta\,(U - U_D) \tag{5.52}$$

where $\epsilon = D_1/D_0$, $\beta = U_A\alpha/\delta$ and U_D is re-scaled according to (5.50). The problem thus depends on three parameters, i.e., β, U_D and ϵ. The conditions at infinity are given by (5.41) where now $K = 1$.

The existence of the two threshold values, i.e., β and U_D, defines, in a general case, three spatial regions according to how the population density in each region compares with these thresholds; see Fig. 5.4. Namely, in the region far to the left (region I), $U > \max(\beta, U_D)$. In the region far to the right (region III), $U < \min(\beta, U_D)$. Correspondingly, in the intermediate region, i.e., region II, we have $\min(\beta, U_D) < U < \max(\beta, U_D)$. Region II disappears in the marginal case $\beta = U_D$.

An essential feature of the above model is that equation (5.40) with population growth given by (5.51) and diffusivity given by (5.52) is linear in each of the regions I, II and III. That makes possible to obtain its analytical solution (Strier et al., 1996).

Apparently, in region I equation (5.40) takes the following form:

$$\epsilon\frac{d^2U}{d\xi^2} + c\frac{dU}{d\xi} + (1 - U) = 0 \tag{5.53}$$

and its biologically meaningful solution (i.e., bounded for $\xi \to -\infty$) is

$$U(\xi) = 1 + A_1 \exp(\lambda_1\xi) \tag{5.54}$$

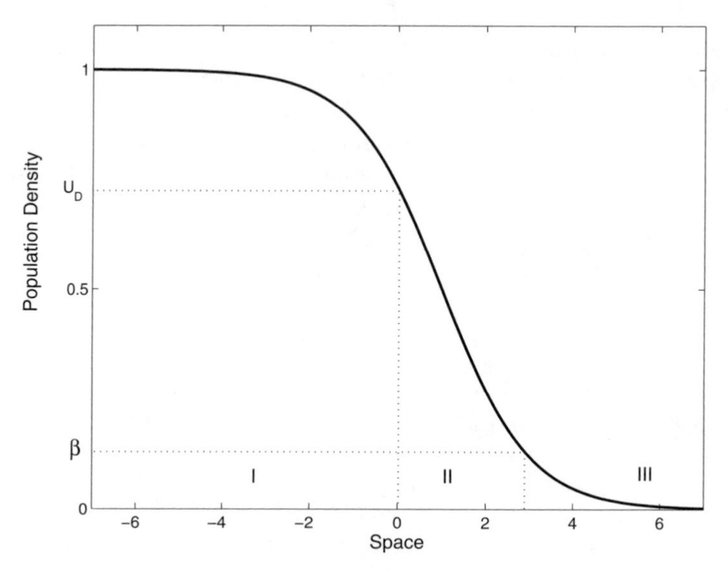

FIGURE 5.4: A sketch of the population density versus space in a traveling population front. Spatial domains I, II and III are defined according to how the population density compares with the two threshold values U_D and β; details are given in the text.

where

$$\lambda_1 = \frac{-c + \sqrt{c^2 + 4\epsilon}}{2\epsilon} . \tag{5.55}$$

In region III, the equation takes the form

$$\frac{d^2U}{d\xi^2} + c\frac{dU}{d\xi} - U = 0 \tag{5.56}$$

and the corresponding solution is

$$U(\xi) = A_3 \exp(\lambda_3 \xi) \tag{5.57}$$

where

$$\lambda_3 = \frac{-c - \sqrt{c^2 + 4}}{2} . \tag{5.58}$$

In the intermediate region, the exact form of the equation depends on the relation between β and U_D. Combining both cases together, the solution takes the following form:

$$U(\xi) = A_2^- \exp(\lambda_2^- \xi) + A_2^+ \exp(\lambda_2^+ \xi) + \theta(U_D - \beta) \tag{5.59}$$

where

$$\lambda_2^{\pm} = \frac{-c \pm \sqrt{c^2 + 4D_i}}{2D_i} \, . \tag{5.60}$$

Here the notation D_i is introduced for convenience. The value of D_i depends on how β and U_D compare so that $D_i = \epsilon$ if $\beta > U_D$ and $D_i = 1$ otherwise.

Constants A_1, A_3 and $A_2^{\pm}$ can be found from the continuity condition at the boundary between different regions, i.e., at the interface where solutions (5.54), (5.57) and (5.59) must match each other. Since we are looking for a solution of the second order differential equation, at the point ξ_β where $U = \beta$ both the density and its first derivative must be continuous:

$$U(\xi_\beta - 0) \;=\; U(\xi_\beta + 0) \, , \qquad \frac{dU(\xi_\beta - 0)}{d\xi} \;=\; \frac{dU(\xi_\beta + 0)}{d\xi} \tag{5.61}$$

where the notations -0 and $+0$ correspond to the limiting values from the left and from the right of ξ_β, respectively.

Similarly, at the point ξ_D where $U = U_D$ we have:

$$U(\xi_D - 0) = U(\xi_D + 0) \, , \tag{5.62}$$

$$D\left(U(\xi_D - 0)\right) \frac{dU(\xi_D - 0)}{d\xi} = D\left(U(\xi_D + 0)\right) \frac{dU(\xi_D + 0)}{d\xi} \, . \tag{5.63}$$

Let us note that the critical coordinates ξ_β and ξ_D are not known either. To determine them, we can use their definition, i.e.,

$$U(\xi_\beta) \;=\; \beta \, , \qquad U(\xi_D) \;=\; U_D \, . \tag{5.64}$$

Hereby, (5.61–5.64) give a system of six equations to be used to obtain seven unknown variables A_1, A_3, $A_2^{\pm}$, ξ_β, ξ_D and c. However, it is readily seen that, since Eq. (5.40) does not contain ξ explicitly, it is invariant with respect to translation. By virtue of translation symmetry, the solution actually depends on the difference $(\xi_\beta - \xi_D)$ rather than on ξ_β and ξ_D separately. It means that one of the unknown coordinates ξ_β and ξ_D can be either kept arbitrary or set to a convenient value. Below we let $\xi_D = 0$.

Special case: $\beta = U_D$. In a general case, calculations appear to be rather cumbersome and tedious. However, if we assume that $\beta = U_D$, the problem is treated analytically much more easily. It should be mentioned that, although this assumption may seem somewhat restrictive, actually, very little is known about the value of U_D in real ecological communities as well as about the mechanisms underlying diffusion density-dependence in general.

In this case, the intermediate region disappears, and the solution matching conditions leads to a system of only three equations, i.e.,

$$U(0 - 0) = U(0 + 0) \, , \tag{5.65}$$

$$D(U(0 - 0)) \frac{dU(0 - 0)}{d\xi} = D(U(0 + 0)) \frac{dU(0 + 0)}{d\xi} , \tag{5.66}$$

and $U(0) = \beta$. From the first and the last of them, we immediately obtain that $A_3 = \beta$ and $A_1 = \beta - 1$; thus the exact solution is given by

$$U(\xi) = 1 - (1-\beta)\exp(\lambda_1 \xi) \quad \text{for} \quad \xi < 0 \; , \tag{5.67}$$

$$U(\xi) = \beta \exp(\lambda_3 \xi) \quad \text{for} \quad \xi > 0 \tag{5.68}$$

where λ_1 and λ_3 are given by (5.55) and (5.58), respectively, and c is yet to be determined.

To obtain the speed, we make use of Eq. (5.66) that now reads as follows:

$$\epsilon(\beta - 1)\lambda_1 = \beta\lambda_3 \; . \tag{5.69}$$

After some standard transformations, from (5.69) we obtain:

$$c = \left[(1-\beta)^2 - \frac{\beta^2}{\epsilon}\right]\left[\left(1 - \beta + \frac{\beta}{\epsilon}\right)(1-\beta)\beta\right]^{-1/2} . \tag{5.70}$$

In the absence of density-dependence, i.e., for $\epsilon = 1$, Eq. (5.70) coincides with (3.94).

Figure 5.5 shows the speed c given by (5.70) versus β for different values of ϵ, the solid curves from left to right correspond to $\epsilon = 0.2,\ 0.33,\ 1.0,\ 3.0$ and 5.0 and the dashed-and-dotted curve shows the limiting case $\epsilon = \infty$. While the unbounded increase of c in vicinity of $\beta = 0$ and $\beta = 1$ is apparently the artifact of the model (see the comments below Eq. (3.94)), in the intermediate region the behavior of the curves looks biologically reasonable. One interesting thing can be immediately inferred from the shape and position of the curves. For the parameters that would result in wave blocking in the case of constant diffusivity, i.e., for $\beta = 0.5$, there will be no blocking when the diffusivity is given by the step function (5.52). In particular, the traveling population front will still correspond to species invasion if higher population density corresponds to higher diffusivity ($\epsilon > 1$), and it will correspond to species retreat otherwise ($\epsilon < 1$). The larger ϵ is, the more invasive is the alien species; in particular, in the large-ϵ limit (cf. the dashed curve) the invasive species can only be blocked if $\beta \approx 1$.

The exact solution given by (5.67), (5.68) and (5.70) is shown in Fig. 5.6 obtained for $\epsilon = 1$ (solid curve), $\epsilon = 5$ (dashed curve) and $\epsilon = 0.2$ (dotted curve). The Allee threshold β was set to a hypothetical value 0.3 so that the corresponding values of speed are 0.87, 1.18 and 0.06, respectively. It is readily seen that the shape of the wave profile appears to be rather sensitive to the value of ϵ; in particular, the width of the front depends on ϵ significantly. This is in a qualitative agreement with the results of field observations showing that the width of the transition region increases considerably when the higher motility mode is "turned on," cf. Section 8.3.

General case. The above assumption about consilience of the two thresholds does not allow to consider the impact of stratified diffusion separately

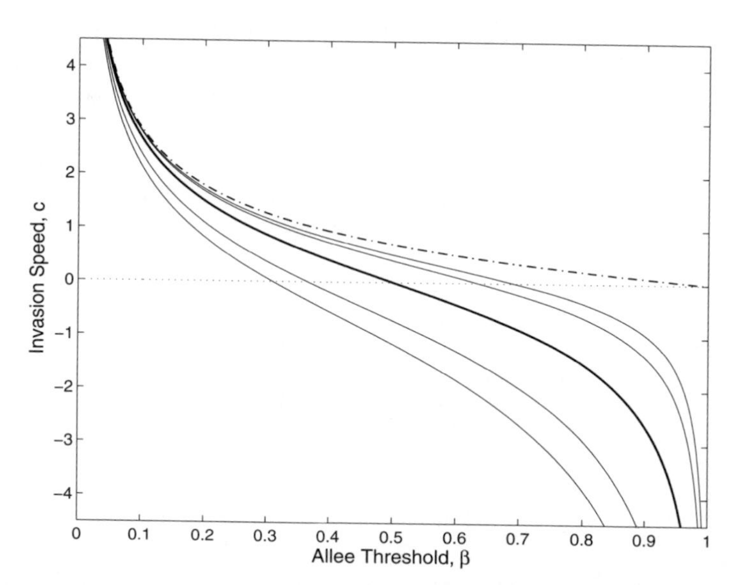

FIGURE 5.5: The speed c of traveling population front versus threshold density β for different values of the diffusivity ratio $\epsilon = D_1/D_0$. Solid curves from left to right correspond to $\epsilon = 0.2,\ 0.33,\ 1,\ 3$ and 5; the dashed-and-dotted curve shows the limiting case $\epsilon = \infty$.

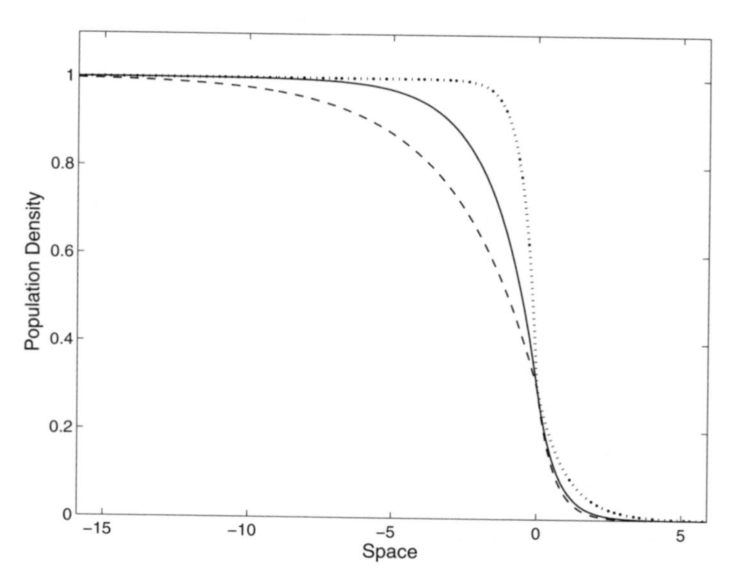

FIGURE 5.6: Exact solution (5.67–5.68) shown for $\epsilon = 1$ (solid curve), $\epsilon = 5$ (dashed curve) and $\epsilon = 0.2$ (dotted curve).

from the impact of the Allee effect. In order to address this issue, now we have to proceed to the general case $\beta \neq U_D$. The system (5.61–5.64) is solved differently depending on how β and U_D compare; thus, two cases should be considered.

Case a: $\beta > U_D$. Taking into account that in this case $\lambda_1 = \lambda_2^+$, from the matching conditions (5.61–5.64), after laborious but standard transformations we obtain the following equations for coefficients A_i:

$$A_1 = \frac{\epsilon\lambda_2^-\lambda_2^+(P-Q) + \lambda_3(\lambda_2^- Q - \lambda_2^+ P)}{(\lambda_2^- - \lambda_2^+)(\epsilon\lambda_2^+ - \lambda_3)} , \tag{5.71}$$

$$A_2^+ = \frac{\lambda_2^+ P(\epsilon\lambda_2^- - \lambda_3)}{(\lambda_2^- - \lambda_2^+)(\epsilon\lambda_2^+ - \lambda_3)} , \qquad A_2^- = \frac{\lambda_2^+ P}{\lambda_2^+ - \lambda_2^-} , \tag{5.72}$$

$$A_3 = \frac{\epsilon\lambda_2^+ P}{\epsilon\lambda_2^+ - \lambda_3} \tag{5.73}$$

where $P = \exp(-\lambda_2^-\xi_\beta)$, $Q = \exp(-\lambda_2^+\xi_\beta)$. The critical coordinate ξ_β can be immediately found from the relation $u(\xi_D) = U_D$ (recalling that $\xi_D = 0$) which now reads as $A_3 = U_D$ and can be explicitly written as follows:

$$\exp(-\lambda_2^-\xi_\beta) = \frac{\epsilon\lambda_2^+ - \lambda_3}{\epsilon\lambda_2^+} U_D . \tag{5.74}$$

Finally, the equation for the wave speed takes the following form:

$$\left(\frac{\epsilon\lambda_2^+ - \lambda_3}{\epsilon\lambda_2^+} U_D\right)^{\lambda_2^+/\lambda_2^-} = \frac{\epsilon\lambda_2^- - \lambda_3}{\epsilon(\lambda_2^- - \lambda_2^+)(\beta - 1) + \epsilon\lambda_2^-} U_D \tag{5.75}$$

where $\lambda_2^\pm$ and λ_3 are given by (5.60) and (5.58), respectively.

Case b: $\beta < U_D$. Now, $\lambda_3 = \lambda_2^-$ and the coefficients are obtained as follows:

$$A_1 = \frac{\lambda_2^- Q}{\epsilon\lambda_1 - \lambda_2^-} , \tag{5.76}$$

$$A_2^+ = \frac{\lambda_2^- Q}{\lambda_2^+ - \lambda_2^-} , \qquad A_2^- = \frac{\lambda_2^- Q[\lambda_2^+ - \epsilon\lambda_1]}{(\lambda_2^+ - \lambda_2^-)[\epsilon\lambda_1 - \lambda_2^-]} , \tag{5.77}$$

$$A_3 = \frac{\lambda_2^+\lambda_2^-(P-Q) - \epsilon\lambda_1(\lambda_2^+ P - \lambda_2^- Q)}{(\lambda_2^+ - \lambda_2^-)[\lambda_2^- - \epsilon\lambda_1]} . \tag{5.78}$$

The value of ξ_β is given as

$$\exp(-\lambda_2^+\xi_\beta) = \frac{\epsilon\lambda_1 - \lambda_2^-}{\lambda_2^-} (U_D - 1) . \tag{5.79}$$

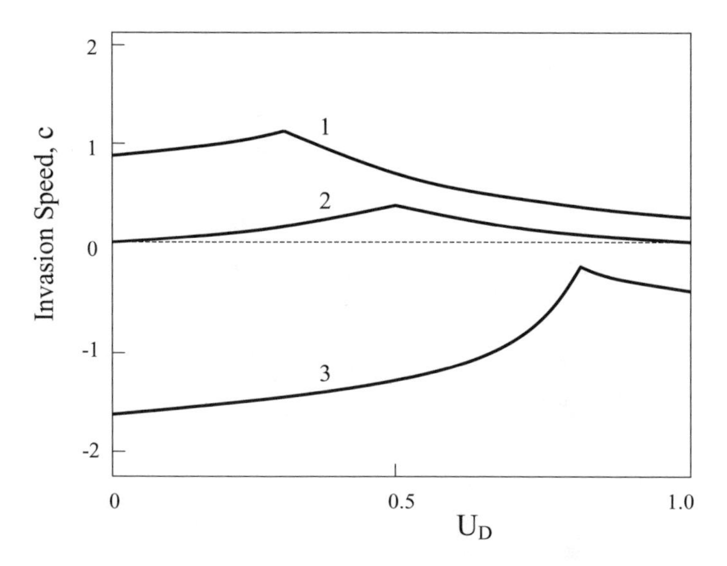

FIGURE 5.7: The speed c of traveling population front versus stratification threshold U_D for $\epsilon = 10$ and $\beta = 0.3$ (curve 1), $\beta - 0.5$ (curve 2), $\beta = 0.8$ (curve 3).

Correspondingly, the wave speed appears as a solution of the following equation:

$$\left(\frac{\epsilon\lambda_1 - \lambda_2^-}{\lambda_2^-} (U_D - 1) \right)^{\lambda_2^- / \lambda_2^+} = \frac{\lambda_2^+ - \epsilon\lambda_1}{(\lambda_2^+ - \lambda_2^-)\beta - \lambda_2^+} (U_D - 1) \quad (5.80)$$

where λ_1 is given by (5.55).

Equations (5.75) and (5.80) are transcendental and can only be solved numerically. The stratified diffusion observed for natural populations indicates that diffusivity tends to increase with the population density, cf. Hengeveld (1989) and Shigesada and Kawasaki (1997); thus, for biological reasons we focus on the case $\epsilon > 1$. Specifically, we choose a hypothetical value $\epsilon = 10$ and consider how the invasion speed changes with U_D for a few selected values of β. The results are shown in Fig. 5.7. An interesting and counterintuitive feature is that the speed appears to depend on U_D in a nonmonotonous way. In particular, for $\beta \leq 0.5$ (cf. curves 1 and 2), as U_D decreases from unity the invasion speed gradually increases and keeps increasing until $U_D = \beta$. Surprisingly, a further decrease in U_D leads to a decrease in the speed. As a result, although the value of speed obtained for $U_D = 0$ is $\sqrt{\epsilon}$ times higher than that obtained for $U_D = 1$ (as it immediately follows from consideration of corresponding populations with density-independent diffusivity), the curves have a hump so that the maximum speed is reached for $U_D = \beta$.

Chapter 6

Models of interacting populations

In the previous chapters, we have considered a variety of models taking into account different features of the population dynamics of invasive species. However, one important factor, namely, the impact of inter-species interactions, has not been addressed yet. Meanwhile, there is considerable evidence that the impact of other species can significantly modify the rate and pattern of exotic species spread. The problem is that the nonlinear models including more than one dynamical variable are usually much more difficult for rigorous mathematical analysis. In particular, only very few exact solutions are known that possess a clear biological meaning. Two of them will be considered below.

The contents of this chapter are based on original papers by Feltham and Chaplain (2000) and Petrovskii et al. (2005a). Exact solutions for a few systems of PDEs were also found by Calogero and Xiaoda (1991); however, since their solutions do not have immediate biological applications, we do not recall their work here.

6.1 Exact solution for a diffusive predator-prey system

Predator-prey relations are among the most common ecological interactions. Moreover, they are probably the most important for ecosystem functioning: it is predation that provides the mechanism of biomass flow through the food web and integrates separate species into a system. Remarkably, the whole field of mathematical ecology began with a study of (spatially homogeneous) population dynamics subject to predator-prey interaction, cf. the classical works by Lotka (1925) and Volterra (1926).

According to a widely accepted approach (see Chapter 2 for details and references), the spatiotemporal dynamics of a predator-prey system can be described by the following equations:

$$\frac{\partial u(x,t)}{\partial t} = D\frac{\partial^2 u}{\partial x^2} + f(u)u - r(u)uv \ , \tag{6.1}$$

$$\frac{\partial v(x,t)}{\partial t} = D\frac{\partial^2 v}{\partial x^2} + \kappa r(u)uv - g(v)v \tag{6.2}$$

where u, v are the densities of prey and predator, respectively, at position x and time t, the function $f(u)$ is the per capita growth rate of the prey, the term $r(u)uv$ stands for predation, κ is the coefficient of food utilization, and $g(v)$ is the per capita mortality rate of the predator. Here, the first term on the right-hand side of Eqs. (6.1–6.2) describes the spatial mixing caused either by self-motion of individuals (Skellam, 1951; Okubo, 1980) or by properties of the environment, e.g., for plankton communities the mixing is attributed to turbulent diffusion (Okubo, 1980). D is the diffusion coefficient; in this section, we assume it to be the same for prey and predator.

We want to mention here that, from the point of ecological applications, the above assumption about equal species diffusivity is not very restrictive. Although predator indeed often has higher diffusivity, a closer inspection of population communities reveals many trophical relations where diffusivity of prey and predator is of the same magnitude. One immediate example is given by a plankton community where spatial mixing takes place mainly due to turbulence which has the same impact on phyto- (prey) and zooplankton (predator). In terrestrial ecosystems, examples may be given by lynx and hare, wolf and deer, etc. In fact, predator's success is often reached not due to a faster motion but due to an optimal foraging strategy while its diffusivity must not be necessarily higher than that of prey.

For different species, functions f, r and g can be of different types. Here we assume that the prey dynamics is subject to the Allee effect so that its per capita growth rate is not a monotonically decreasing function of the prey density but possesses a local maximum. We will focus on the strong Allee effect when the prey growth rate becomes negative for $0 < u < u_A$ where u_A is a certain threshold density. In this case, the standard parameterization is as follows:

$$f(u) = \omega(u - u_A)(K - u); \tag{6.3}$$

see Section 1.2 for details and references.

Regarding the per capita predator mortality, we assume that it is described by the following function:

$$g(v) = M + d_0 v^n \tag{6.4}$$

where M, d_0 and n are positive parameters. Function $g(v)$ gives the so-called "closure term" because it is supposed to not only describe the processes going on inside the predator population such as natural (linear) mortality, competition, possibly cannibalism, etc., but also virtually take into account the impact of higher predators which are not included into the model explicitly (Steele and Henderson, 1992a). Different authors consider various functional forms for the closure term, particularly, different n, e.g., see Edwards and Yool (2000) and references therein. It should be mentioned, however, that the accuracy of ecological observations is usually rather low so that it is not often possible to give a reliable estimate of n. For the sake of analytical tractability, we restrict our consideration to the case $n = 2$.

Finally, we assume that the predator shows a linear response to prey according to the classical Lotka–Volterra model, i.e., $r(u) = \eta = const.$ Then, Eqs. (6.1–6.2) with (6.3), (6.4) take the following form:

$$\frac{\partial u(x,t)}{\partial t} = D\frac{\partial^2 u}{\partial x^2} + \omega u(u - u_A)(K - u) - \eta uv \ , \tag{6.5}$$

$$\frac{\partial v(x,t)}{\partial t} = D\frac{\partial^2 v}{\partial x^2} + \kappa\eta uv - Mv - d_0 v^3 \ . \tag{6.6}$$

Introducing dimensionless variables

$$\tilde{u} = \frac{u}{K} \ , \quad \tilde{v} = \frac{\eta v}{\omega K^2} \ , \quad \tilde{x} = x\sqrt{\frac{\omega K^2}{D}} \ , \quad \tilde{t} = t\omega K^2 \ , \tag{6.7}$$

and omitting tildes further on for notation simplicity, from Eqs. (6.5–6.6) we obtain:

$$u_t = u_{xx} - \beta u + (\beta + 1)u^2 - u^3 - uv \ , \tag{6.8}$$

$$v_t = v_{xx} + kuv - mv - \delta v^3 \tag{6.9}$$

where $\beta = u_A K^{-1}$, $k = \kappa\eta(\omega K)^{-1}$, $m = M(\omega K^2)^{-1}$, $\delta = d_0\omega K^2\eta^{-2}$ are positive dimensionless parameters, subscripts x and t stand for the partial derivatives with respect to dimensionless space and time, respectively. We consider Eqs. (6.8–6.9) in an infinite space, $-\infty < x < \infty$, and for $t > 0$. Functions $u(x,t)$, $v(x,t)$ are assumed to be bounded for $x \to \pm\infty$. To make the problem complete, Eqs. (6.8–6.9) should be provided with the initial conditions $u(x,0) = u_0(x)$ and $v(x,0) = v_0(x)$; the form of $u_0(x)$, $v_0(x)$ will be specified below.

In the previous chapters, we have seen that exact solutions of nonlinear PDEs are usually ad hoc solutions obtained either for a specific form of nonlinearity or for certain restrictions on parameter values, although it is difficult to say whether that stems from immaturity of contemporary theory of nonlinear partial differential equations or reflects certain intrinsic symmetries of given PDEs. In the rest of this paper, for the sake of analytical tractability, we assume the following relations between the equation parameters:

$$m = \beta \ , \tag{6.10}$$

$$k + \frac{1}{\sqrt{\delta}} = \beta + 1 \ . \tag{6.11}$$

The origin and the meaning of constraints (6.10–6.11) will become clear later.

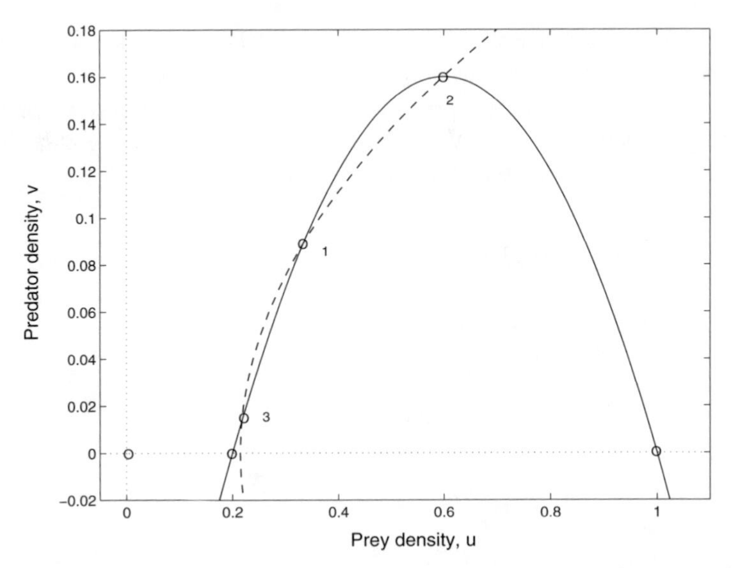

FIGURE 6.1: Zero-isoclines for the predator-prey model (6.12–6.13). The circles show the steady states of the system; numbering is explained in the text.

6.1.1 * Properties of the local system

Before proceeding to the analysis of system (6.8–6.9), we are going to give a brief insight into the properties of the corresponding spatially homogeneous system:

$$u_t = -\beta u + (\beta + 1)u^2 - u^3 - uv \ , \tag{6.12}$$

$$v_t = kuv - mv - \delta v^3 \ . \tag{6.13}$$

The question of primary importance is the existence of steady states which, as usual in the case of autonomous systems of second order, are given by the intersection points of the zero-isoclines of the system (see Fig. 6.1). It is readily seen that Eqs. (6.12–6.13) always have the trivial steady state $(0,0)$ corresponding to species extinction and the two "prey only" states $(\beta,0)$, $(1,0)$. As for the existence and the number of the steady states in the interior of the first quadrant, i.e., in $\mathbf{R}_+^2 = \{(u,v) \mid u > 0,\ v > 0\}$, it is somewhat less obvious. The following theorem addresses this issue.

Theorem 6.1. *Let conditions (6.10–6.11) be satisfied and, additionally, $k > 2\sqrt{m}$. Then the number of steady states of system (6.12–6.13) inside $\mathbf{R}_+^2$ depends on the sign of the following quantity:*

$$h = \frac{k}{\sqrt{\delta}} - m \ . \tag{6.14}$$

Namely, there are two steady states in the case $h < 0$ and three steady states in the case $h > 0$.

Proof. Obviously, inside $\mathbf{R}^2_+$ the isoclines of the system (6.12–6.13) are given by the following equations:

$$v = -\beta + (\beta + 1)u - u^2 , \tag{6.15}$$

$$u = \frac{m}{k} + \frac{\delta}{k}v^2 \tag{6.16}$$

(see Fig. 6.1). For any steady state $(\bar{u}, \bar{v})$, $\bar{u}$ and $\bar{v}$ are a solution of system (6.15–6.16). Having substituted (6.15) into (6.16), we obtain the following equation for $\bar{u}$

$$\begin{aligned} \frac{k}{\delta}\bar{u} - \frac{m}{\delta} &= \beta^2 - 2\beta(\beta + 1)\bar{u} \\ &+ \left(2\beta + (\beta + 1)^2\right)\bar{u}^2 - 2(\beta + 1)\bar{u}^3 + \bar{u}^4 \end{aligned} \tag{6.17}$$

which, taking into account conditions (6.10–6.11), after standard although tedious algebraic transformations can be written in the following form:

$$\left(\bar{u}^2 - k\bar{u} + m\right)\left[\bar{u}^2 - \left(k + \frac{2}{\sqrt{\delta}}\right)\bar{u} + \left(m + \frac{1}{\delta}\right)\right] = 0 . \tag{6.18}$$

Thus, possible stationary values $\bar{u}$ are given by

$$\bar{u}_{1,2} = \frac{k}{2} \pm \sqrt{\left(\frac{k}{2}\right)^2 - m} \tag{6.19}$$

and

$$\bar{u}_{3,4} = \frac{1}{2}\left[\left(k + \frac{2}{\sqrt{\delta}}\right) \pm \sqrt{\left(k + \frac{2}{\sqrt{\delta}}\right)^2 - 4\left(m + \frac{1}{\delta}\right)}\,\right] . \tag{6.20}$$

It is readily seen that, under the assumptions of Theorem 6.1, all $\bar{u}_i > 0,\ i = 1, \ldots, 4$.

The corresponding stationary values $\bar{v}_i$ can be obtained substituting Eqs. (6.19) and (6.20) into (6.15). However, we use another approach which appears to be less cumbersome. Having substituted (6.16) to (6.15) and taking into account (6.10–6.11), we obtain the equation for $\bar{v}$ which, after transformations, can be written in the following form:

$$\left(\delta\bar{v}^2 - k\sqrt{\delta}\bar{v} + m\right)\left[\delta\bar{v}^2 + k\sqrt{\delta}\bar{v} + \left(m - \frac{k}{\sqrt{\delta}}\right)\right] = 0 . \tag{6.21}$$

Thus, the stationary values $\bar{v}$ are given by

$$\bar{v}_{1,2} = \frac{1}{\sqrt{\delta}}\left[\frac{k}{2} \pm \sqrt{\left(\frac{k}{2}\right)^2 - m}\,\right] \tag{6.22}$$

and

$$\bar{v}_{3,4} = \frac{1}{\sqrt{\delta}} \left[-\frac{k}{2} \pm \sqrt{\left(\frac{k^2}{4} - m\right) + \frac{k}{\sqrt{\delta}}} \right] \tag{6.23}$$

where plus and minus correspond to $\bar{v}_3$ and $\bar{v}_4$, respectively.

Under the assumptions of Theorem 6.1, $v_{1,2} > 0$ and $\bar{v}_4 < 0$. Thus, the steady states $(\bar{u}_1, \bar{v}_1)$, $(\bar{u}_2, \bar{v}_2)$ lie inside $\mathbf{R}^2_+$ and $(\bar{u}_4, \bar{v}_4)$ lies outside $\mathbf{R}^2_+$. The only remaining question is about the sign of $\bar{v}_3$ and the corresponding steady state $(\bar{u}_3, \bar{v}_3)$.

Obviously, the following inequality

$$\bar{v}_3 = \frac{1}{\sqrt{\delta}} \left[-\frac{k}{2} + \sqrt{\left(\frac{k^2}{4} - m\right) + \frac{k}{\sqrt{\delta}}} \right] > 0 \tag{6.24}$$

is equivalent to

$$\frac{k}{\sqrt{\delta}} - m > 0 \, . \tag{6.25}$$

Thus, the steady state $(\bar{u}_3, \bar{v}_3)$ lies inside $\mathbf{R}^2_+$ if inequality (6.25) is true and outside $\mathbf{R}^2_+$ otherwise. That proves Theorem 6.1.

Another important point is the steady states' stability. Taking into account that the stationary values $\bar{u}$, $\bar{v}$ appear as the solutions of fourth-order algebraic equations, a thorough investigation of this issue is very difficult. However, due to the specifics of the exact solution which will be presented below, for the goals of this paper it seems possible to restrict our consideration to the stability of only two steady states, i.e., $(0,0)$ and $(\bar{u}_2, \bar{v}_2)$ where subscript "2" corresponds to plus in Eqs. (6.19) and (6.22). It is straightforward to see that $(0,0)$ is always stable. As for $(\bar{u}_2, \bar{v}_2)$, a sufficient condition of its stability is given by the following theorem.

Theorem 6.2. *Let conditions (6.10–6.11) be satisfied and*

$$k > 2\sqrt{m + \frac{1}{4\delta}} \, . \tag{6.26}$$

Then the steady state $(\bar{u}_2, \bar{v}_2)$ is stable.

Proof. The conclusion of the theorem nearly immediately follows from a more general fact that a steady state is stable when it arises as an intersection point of decreasing zero-isocline for prey (i.e., originated from the equation for prey) and increasing zero-isocline for predator. In terms of system (6.12–6.13), it means that $(\bar{u}_2, \bar{v}_2)$ is stable when it is situated on the right of the hump of curve 1, cf. Fig. 6.1. Thus, all we need is to compare $\bar{u}_2$ with the position of the hump. Under conditions (6.10–6.11), $\bar{u}_2$ is given by Eq. (6.19) and the maximum of the isocline for predator is situated at $(\beta+1)/2$. Then, taking into account (6.11), the steady state $(\bar{u}_2, \bar{v}_2)$ is stable for

$$\frac{k}{2} + \sqrt{\left(\frac{k}{2}\right)^2 - m} > \frac{1}{2}\left(k + \frac{1}{\sqrt{\delta}}\right), \tag{6.27}$$

which is equivalent to assumption (6.26). That proves the theorem.

6.1.2 Exact solution and its properties

The main result of this section is given by the following theorem:

Theorem 6.3. *Under constraints (6.10–6.11) and $k > 2\sqrt{m}$, the system (6.8–6.9) has the following exact solution:*

$$u(x,t) = \frac{\bar{u}_1 \exp(\lambda_1\xi_1) + \bar{u}_2 \exp(\lambda_2\xi_2)}{1 + \exp(\lambda_1\xi_1) + \exp(\lambda_2\xi_2)}, \tag{6.28}$$

$$v(x,t) = \frac{\bar{v}_1 \exp(\lambda_1\xi_1) + \bar{v}_2 \exp(\lambda_2\xi_2)}{1 + \exp(\lambda_1\xi_1) + \exp(\lambda_2\xi_2)} \tag{6.29}$$

where $\bar{u}_{1,2}$ and $\bar{v}_{1,2}$ are the steady states of the system (6.12–6.13) given by Eqs. (6.19) and (6.22), respectively, $\xi_1 = x - n_1 t + \phi_1$, $\xi_2 = x - n_2 t + \phi_2$,

$$\lambda_{1,2} = \frac{1}{2\sqrt{2}}\left(k \pm \sqrt{k^2 - 4m}\right), \quad n_i = \sqrt{2}k - 3\lambda_i, \quad i = 1, 2, \tag{6.30}$$

and $\phi_{1,2}$ are arbitrary constants.

Theorem 6.3 can be proved immediately substituting (6.28–6.29) into Eqs. (6.8–6.9). A simple way to arrive at solution (6.28–6.29) is also shown below. A formal approach based on an appropriate change of variables that leads to (6.28–6.29) and also helps to understand the origin of the relations (6.10–6.11) is given in the next section.

Since the solution (6.28–6.29) is expected to have a variety of ecological/biological applications (some of them will be discussed in Chapter 8), we are going to have a closer look at its properties. The form of (6.28–6.29) suggests that it describes propagation of two waves traveling with the speeds n_1 and n_2, correspondingly; see Fig. 6.2. Assuming, without any loss of generality, that $\lambda_1 < \lambda_2$ (i.e., choosing minus for λ_1 and plus for λ_2 in Eqs. (6.30)), we immediately obtain that

$$n_1 = \frac{1}{2\sqrt{2}}\left(k + 3\sqrt{k^2 - 4m}\right), \tag{6.31}$$

$$n_2 = \frac{1}{2\sqrt{2}}\left(k - 3\sqrt{k^2 - 4m}\right). \tag{6.32}$$

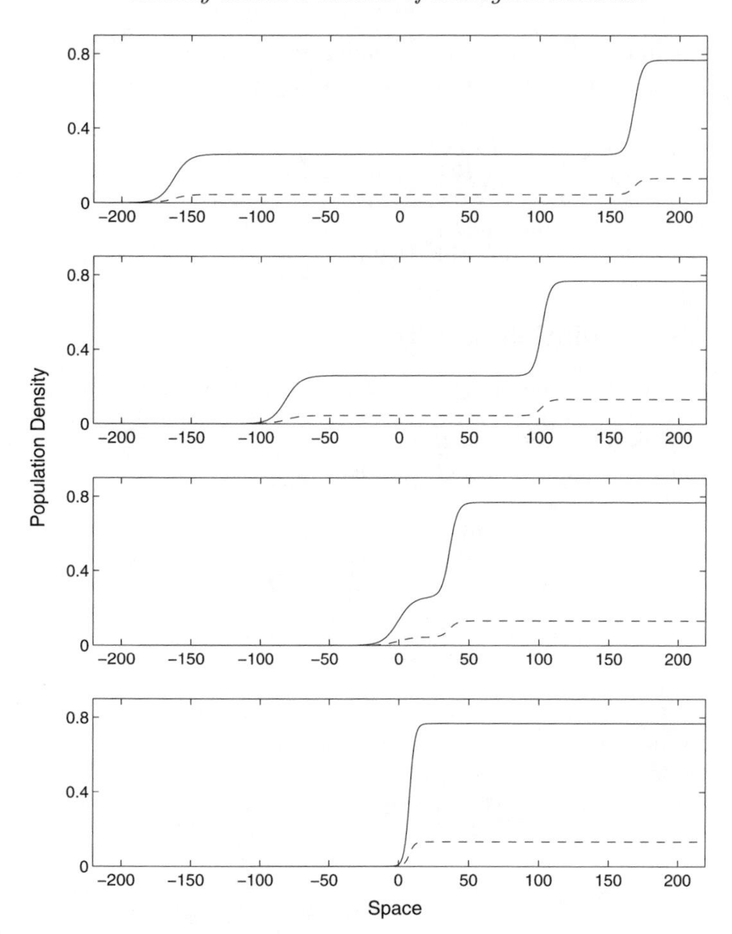

FIGURE 6.2: Exact solution (6.28–6.29) shown at equidistant moments, from top to bottom, $t = 0$, $t = 90$, $t = 180$ and $t = 270$. Solid and dashed curves show prey and predator density, respectively. Formation of a population front (at the bottom) is preceded by a decay of unstable spatially homogeneous distribution (at the top, middle). Parameters are: $m = \beta = 0.2$, $\delta = 34$, $k = 1.03$, $\phi_1 = 30$, $\phi_2 = -30$.

Obviously, $n_1 > n_2$ and n_1 is always positive while n_2 can be either positive or negative depending on the parameter values:

$$n_2 < 0 \quad \text{for} \quad k > 3\sqrt{\frac{m}{2}} \quad \text{and} \quad n_2 > 0 \quad \text{for} \quad k < 3\sqrt{\frac{m}{2}} \,. \tag{6.33}$$

In order to get an insight into the nature of the waves, let us consider the case when, for certain x and t, $\lambda_1\xi_1 \simeq 1$ or even $\lambda_1\xi_1 \gg 1$ while $\lambda_2\xi_2$ is negative and large. For sufficiently small t, it can always be achieved by a proper choice of ϕ_1 and ϕ_2. Then, in this domain,

$$u(x,t) \simeq \frac{\bar{u}_1 \exp(\lambda_1\xi_1)}{1+\exp(\lambda_1\xi_1)}\,, \quad v(x,t) \simeq \frac{\bar{v}_1 \exp(\lambda_1\xi_1)}{1+\exp(\lambda_1\xi_1)}\,. \tag{6.34}$$

Thus, the wave propagating with the speed n_1 is a traveling front connecting the steady states $(0,0)$ and $(\bar{u}_1, \bar{v}_1)$.

Owing to $n_1 > n_2$, at any fixed point x variable ξ_1 decreases at a higher rate than ξ_2 and, regardless of the values of ϕ_1 and ϕ_2, for sufficiently large time, the solution (6.28–6.29) can be approximated as follows:

$$u(x,t) \simeq \frac{\bar{u}_2 \exp(\lambda_2\xi_2)}{1+\exp(\lambda_2\xi_2)}\,, \quad v(x,t) \simeq \frac{\bar{v}_2 \exp(\lambda_2\xi_2)}{1+\exp(\lambda_2\xi_2)}\,. \tag{6.35}$$

Therefore, in the large-time limit, the solution (6.28–6.29) describes a traveling front connecting the states $(0,0)$ and $(\bar{u}_2, \bar{v}_2)$ and propagating with the speed n_2. Importantly, the direction of propagation depends on parameter values, cf. (6.33) and see Figs. 6.3 and 6.4.

Curiously, the actual dynamics described by (6.28–6.29) is not exhausted by the two traveling fronts propagating with speeds n_1 and n_2. Considering $\lambda_1\xi_1 \gg 1$, in the crossover region where $\lambda_1\xi_1$ and $\lambda_2\xi_2$ are of the same magnitude we obtain from Eqs. (6.28–6.29):

$$u(x,t) \simeq \frac{\bar{u}_1 + \bar{u}_2 \exp(\lambda_2\xi_2 - \lambda_1\xi_1)}{1+\exp(\lambda_2\xi_2 - \lambda_1\xi_1)}\,, \tag{6.36}$$

$$v(x,t) \simeq \frac{\bar{v}_1 + \bar{v}_2 \exp(\lambda_2\xi_2 - \lambda_1\xi_1)}{1+\exp(\lambda_2\xi_2 - \lambda_1\xi_1)}\,. \tag{6.37}$$

Since

$$\lambda_2\xi_2 - \lambda_1\xi_1 = (\lambda_2 - \lambda_1)\left(x + \frac{k}{\sqrt{2}}\,t + \phi\right) \tag{6.38}$$

where $\phi = (\lambda_2\phi_2 - \lambda_1\phi_1)/(\lambda_2 - \lambda_1)$, the solution in the crossover region apparently behaves as a traveling wave connecting the states $(\bar{u}_1, \bar{v}_1)$ and $(\bar{u}_2, \bar{v}_2)$ and propagating with the speed $-k/\sqrt{2}$.

Thus, in general, the propagation of the traveling front (6.35) can be preceded by the propagation of the "partial" fronts (6.34) and (6.36–6.37) (see Fig. 6.2). However, in the case that ϕ_2 is significantly larger than ϕ_1, the traveling fronts (6.34) and (6.36–6.37) may be never seen explicitly and the exact solution (6.28–6.29) is well approximated by (6.35) for any $t > 0$, cf. Figs. 6.3 and 6.4.

The initial conditions corresponding to the exact solution (6.28–6.29) can be immediately obtained from (6.28–6.29) letting $t = 0$. An important point is, however, that the meaning of solution (6.28–6.29) is not restricted to this specific case. Numerical simulations of system (6.8–6.9) show that the profile

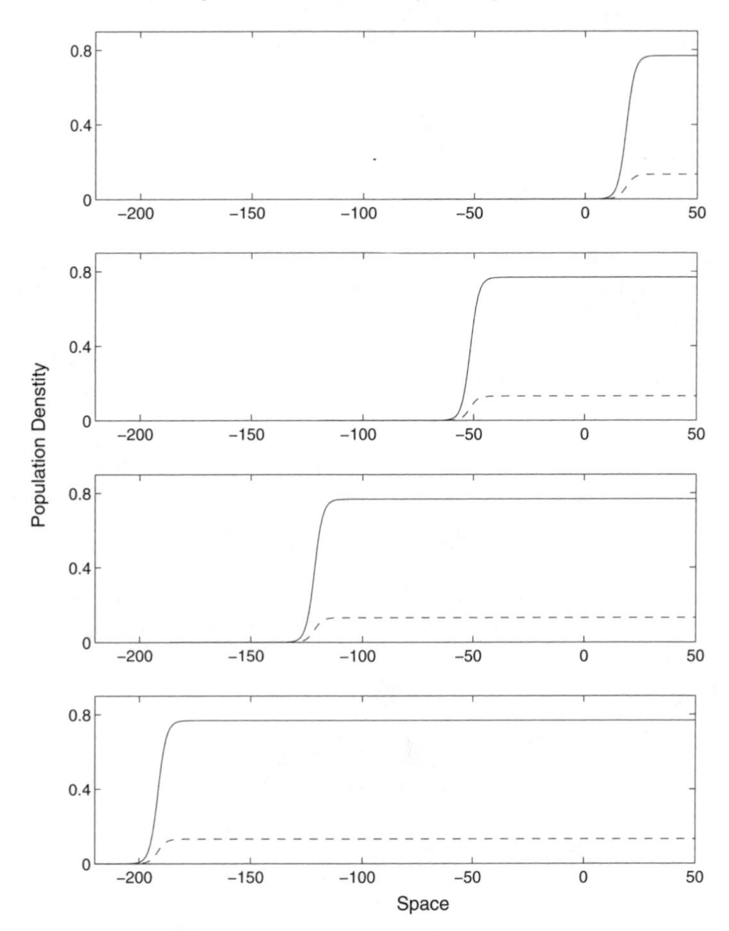

FIGURE 6.3: Exact solution (6.28–6.29) shown at equidistant moments, from top to bottom, $t = 0$, $t = 400$, $t = 800$ and $t = 1200$. Solid and dashed curves show prey and predator density, respectively. Front propagation corresponds to species invasion. Parameters are: $m = \beta = 0.2$, $\delta = 34$, $k = 1.03$, $\phi_1 = -20$, $\phi_2 = -10$.

described by (6.28–6.29) actually appears as a result of convergence for initial conditions from a wide class. Fig. 6.5 gives an example of such convergence obtained in the case that the initial conditions for Eqs. (6.8–6.9) are chosen as piecewise-constant functions, i.e., $u(x,0) = 0$, $v(x,0) = 0$ for $x < 0$ and $u(x,0) = \bar{u}_2$, $v(x,0) = \bar{v}_2$ for $x > 0$.

In the large-time limit, the solution describes a traveling front connecting the state $(0,0)$ which is always stable to the state $(\bar{u}_2, \bar{v}_2)$ which is stable under the assumptions of Theorem 6.2. It should be mentioned here that Theorem 6.2 gives only a sufficient condition of stability, not a necessary one. Although

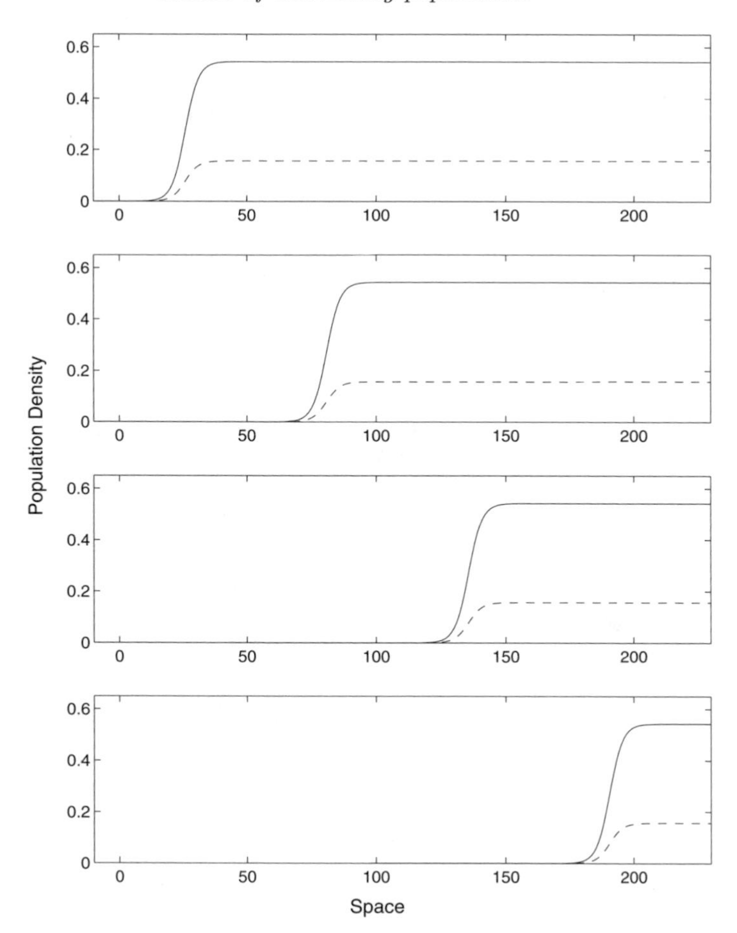

FIGURE 6.4: Exact solution (6.28–6.29) shown at equidistant moments, from top to bottom, $t = 0$, $t = 400$, $t = 800$ and $t = 1200$. Solid and dashed curves show prey and predator density, respectively. Front propagation corresponds to species retreat. Parameters are: $m = \beta = 0.2$, $\delta = 12$, $k = 0.91$, $\phi_1 = -20$, $\phi_2 = -10$.

we cannot prove stability of $(\bar{u}_2, \bar{v}_2)$ under less restrictive assumptions, the results of numerical solution of Eqs. (6.12–6.13) show that the actual range of its stability is wider than the one given by Theorem 6.2. Also, the results of numerical simulations show that the state $(\bar{u}_1, \bar{v}_1)$ is unstable in a wide range of parameter values. These results on the steady states' stability helps to better understand the transient dynamics described by the solution (6.28–6.29) at small times: the partial traveling fronts (6.34) and (6.36–6.37) propagating toward each other correspond to a decay of the quasi-homogeneous species distribution at the unstable level $(\bar{u}_1, \bar{v}_1)$ to the stable homogeneous distribu-

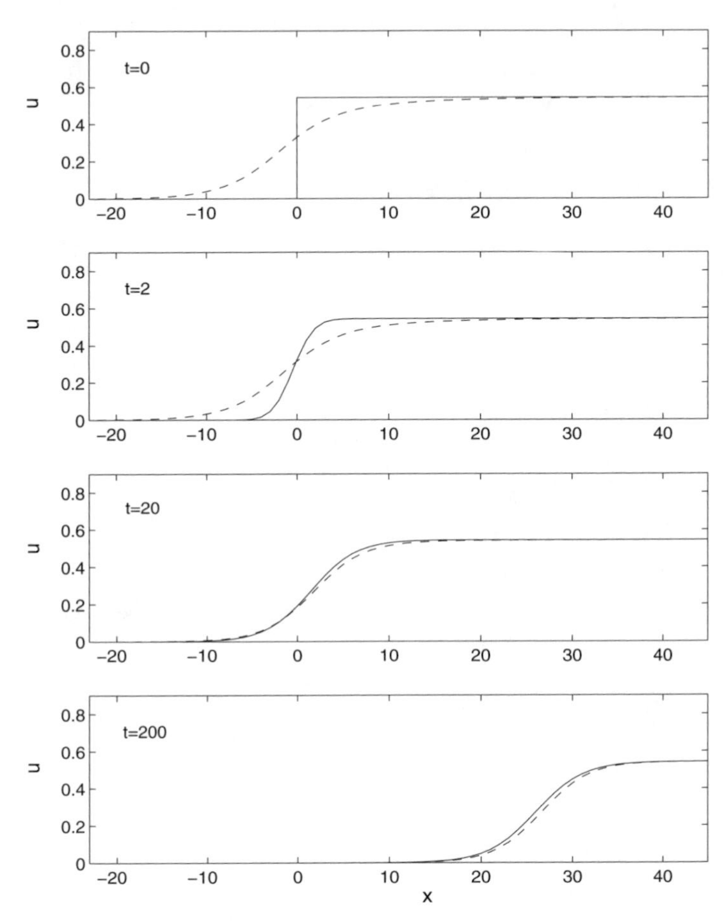

FIGURE 6.5: Convergence of a piecewise constant initial species distribution to the traveling wave profile described by (6.28–6.29). Solid curve shows the results of numerical integration of the system (6.8–6.9); dashed curve shows the exact solution for parameters $m = 0.2$, $\delta = 12$, $k = 0.91$, $\phi_1 = 0$, $\phi_2 = 0.3$. Only prey density is shown. Predator density exhibits similar behavior.

tions at $(0, 0)$ and $(\bar{u}_2, \bar{v}_2)$.

In conclusion to this section, let us have a closer look at the relations (6.10–6.11) between the interaction parameters. In spite of their rather special form, they can be given a clear ecological interpretation. Consider first (6.10). Both β and m have the same meaning. They give linear per capita mortality rate of prey and predator, respectively, in the case that all density-dependence phenomena can be neglected, e.g., in the case of low population density.

To reveal the meaning of (6.11), let us look at the reaction terms in the right-hand side of Eqs. (6.8–6.9) from the point of (bio)mass vertical flow through the food web. Parameter k then quantifies the mass flow from the lower level (prey) to the upper level (predator) while δ quantifies the mass flow from predator to the higher trophical levels that are virtually taken into account by the last term in (6.9), cf. the lines after Eq. (6.4). Apparently, the larger k is, the larger is the mass inflow, and the smaller δ is, the smaller is the mass outflow. It means that the mass kept at the predator level can be quantified by the expression $(k + \delta^{-\nu})$ where ν is a positive parameter. On the other hand, the actual biomass production at the prey level is described by $(\beta + 1)u^2$. Constraint (6.11) thus means that, up to the exact value of ν, the rate of biomass production at the lower level must be consistent with the rate of biomass assimilation at the upper level of this simple trophic web. It may mean that there should be a certain similarity between the populations of prey and predator regarding their response to vertical biomass flow.

Altogether, along with the assumption of equal diffusion coefficients and equal mortality rates, it means that the exact solution (6.28–6.29) describes the dynamics of a population system where prey and predator are in some sense similar, e.g., belong to the same taxonomic group. Indeed, one can observe that under constraints (6.10–6.11) variables u and v become proportional to each other, $v = u/\sqrt{\delta}$. The system (6.8–6.9) is then virtually reduced to one equation:

$$u_t = u_{xx} - \beta u + \left(\beta + 1 - \frac{1}{\sqrt{\delta}}\right) u^2 - u^3 \; . \tag{6.39}$$

Thus, solution (6.28–6.29) gives an extension of the Kawahara–Tanaka (1983) solution to the case of a system of two interacting species. We want to emphasize, however, that this extension is nontrivial. In particular, the solution (6.28–6.29) contains a new mechanism of invasion species blocking which is impossible in the single-species case. Details of this mechanism and its ecological relevance will be discussed in Section 8.2.

6.1.3 * Formal derivation of the exact solution

The formal way to obtain the exact solution (6.28–6.29) is given by the extension of the procedure described in Section 3.2 to the case of two diffusion-reaction equations.

Let us introduce new variables $z(x,t)$ and $w(x,t)$ by means of the following equations:

$$u(x,t) = \mu \frac{z_x}{z + w + \sigma} \; , \quad v(x,t) = \gamma \frac{w_x}{z + w + \sigma} \tag{6.40}$$

where μ and γ are parameters, and σ is a constant included into the denominator of (6.40) in order to avoid singularities because, for biological reasons, we are primarily interested in bounded solutions of system (6.8–6.9). If we restrict our analysis to the case that $z+w$ is semi-bounded, i.e., either $z+w \geq \bar{c}$

or $z+w \le \bar{c}$ for $\forall\, x, t$, then σ can have a nearly arbitrary value with the only restriction $\sigma > -\bar{c}$ or $\sigma < -\bar{c}$, respectively.

Substitution of Eqs. (6.40) into (6.8) leads to the following equation:

$$\begin{aligned}
\left[2\mu z_x(z_x+w_x)^2 - \mu^3 z_x^3\right](z+w+\sigma)^{-3} \\
+[\mu z_x(z_t+w_t) - 2\mu z_{xx}(z_x+w_x) - \mu z_x(z_{xx}+w_{xx}) \\
+(\beta+1)\mu^2 z_x^2 - \mu\gamma z_x w_x](z+w+\sigma)^{-2} \\
+\left[\mu z_{xxx} + \beta\mu z_x - \mu z_{xt}\right](z+w+\sigma)^{-1} = 0 \ .
\end{aligned} \tag{6.41}$$

Since σ is (nearly) arbitrary and functions $(z+w)^j$ are linearly independent for different j (except for the trivial case $z+w \equiv const$), Eq. (6.41) holds if and only if the expressions in the square brackets equal zero identically. Thus, after some obvious transformations we arrive at the following system:

$$0 = 2\mu z_x\left[(z_x+w_x)^2 - \frac{\mu^2}{2}z_x^2\right], \tag{6.42}$$

$$z_t + w_t = 2\frac{z_{xx}}{z_x}(z_x+w_x) + (z_{xx}+w_{xx}) - (\beta+1)\mu z_x + \gamma w_x \ , \tag{6.43}$$

$$z_{xt} = z_{xxx} + \beta z_x \tag{6.44}$$

(assuming that $z_x \neq 0$).

Similarly, substituting Eqs. (6.40) into (6.9) and assuming $w_x \neq 0$, we obtain the following equations:

$$0 = 2\gamma w_x\left[(z_x+w_x)^2 - \frac{\delta\mu^2}{2}w_x^2\right], \tag{6.45}$$

$$z_t + w_t = 2\frac{w_{xx}}{w_x}(z_x+w_x) + (z_{xx}+w_{xx}) - k\mu z_x \ , \tag{6.46}$$

$$w_{xt} = w_{xxx} - m w_x \ . \tag{6.47}$$

The idea of the further analysis is that the system (6.42–6.47) is over-determined and its consistency may only take place under certain constraints on the parameter values. In particular, Eqs. (6.42) and (6.45) are equivalent to

$$z_x + w_x = \pm\frac{\mu}{\sqrt{2}}z_x \tag{6.48}$$

and

$$z_x + w_x = \pm\gamma\sqrt{\frac{\delta}{2}}w_x \ , \tag{6.49}$$

correspondingly.

If we choose plus in both of the above equations, Eqs. (6.48–6.49) take the form

$$w_x = -\left(1 - \frac{\mu}{\sqrt{2}}\right)z_x \tag{6.50}$$

and

$$z_x = -\left(1 - \gamma\sqrt{\frac{\delta}{2}}\right) w_x \ , \tag{6.51}$$

respectively, which is consistent only in case of the following relation between μ and γ:

$$\left(1 - \frac{\mu}{\sqrt{2}}\right)\left(1 - \gamma\sqrt{\frac{\delta}{2}}\right) = 1 \ . \tag{6.52}$$

Next, Eqs. (6.43) and (6.46) become equivalent when their right-hand sides coincide. Taking into account (6.50–6.51), that leads to the following equation:

$$\frac{3}{\sqrt{2}}\mu z_{xx} - \left[(\beta+1)\mu + \gamma\left(1 - \frac{\mu}{\sqrt{2}}\right)\right] z_x$$
$$= \left[1 - (\sqrt{2}\delta\gamma + 1)\left(1 - \frac{\mu}{\sqrt{2}}\right)\right] z_{xx} \quad - \quad k\mu z_x \ . \tag{6.53}$$

Obviously, Eq. (6.53) becomes trivial under the following constraints on the parameter values:

$$\frac{3}{\sqrt{2}}\mu = 1 - (\sqrt{2}\delta\gamma + 1)\left(1 - \frac{\mu}{\sqrt{2}}\right), \tag{6.54}$$

$$k\mu = (\beta+1)\mu + \gamma\left(1 - \frac{\mu}{\sqrt{2}}\right). \tag{6.55}$$

It is readily seen that under condition (6.11) each of Eqs. (6.52), (6.54) and (6.55) is equivalent to the following one:

$$\frac{\gamma}{\mu}\left(1 - \frac{\mu}{\sqrt{2}}\right) = -\frac{1}{\sqrt{\delta}} \ . \tag{6.56}$$

The system (6.42–6.47) is now reduced to only three equations:

$$z_t + w_t = \frac{3}{\sqrt{2}}\mu z_{xx} - \left[(\beta+1)\mu + \gamma\left(1 - \frac{\mu}{\sqrt{2}}\right)\right] z_x \ , \tag{6.57}$$

$$z_{xt} = z_{xxx} - \beta z_x \ , \tag{6.58}$$

$$w_{xt} = w_{xxx} - m w_x \ . \tag{6.59}$$

The number of equations still exceeds the number of variables. However, under condition (6.10) Eqs. (6.58) and (6.59) become identical and the system is reduced to only two equations.

Differentiating Eq. (6.57) with respect to x, substituting (6.58) and (6.59) and taking into account (6.50–6.51) and (6.56), we obtain the following equation for $z(x,t)$:

$$z_{xxx} - \frac{k}{\sqrt{2}} z_{xx} + \frac{m}{2} z_x = 0 \ . \tag{6.60}$$

Eq. (6.60) is linear and its general solution has the following form:

$$z(x,t) = f_0(t) + f_1(t)e^{\lambda_1 x} + f_2(t)e^{\lambda_2 x} \tag{6.61}$$

where the functions $f_0,\ f_1,\ f_2$ still need to be determined and the eigenvalues $\lambda_{1,2}$ are the solutions of the following equation:

$$\lambda^2 - \frac{k}{\sqrt{2}}\lambda + \frac{m}{2} = 0\ , \tag{6.62}$$

so that

$$\lambda_{1,2} = \frac{1}{2\sqrt{2}}\left(k \pm \sqrt{k^2 - 4m}\right). \tag{6.63}$$

Taking into account Eqs. (6.50) and (6.61), we obtain:

$$w(x,t) = g_0(t) - \left(1 - \frac{\mu}{\sqrt{2}}\right)\left[f_1(t)e^{\lambda_1 x} + f_2(t)e^{\lambda_2 x}\right] \tag{6.64}$$

where $g_0(t)$ is a certain function.

To find the functions $f_0,\ f_1,\ f_2,\ g_0$, we substitute (6.61) and (6.64) into Eq. (6.57):

$$\begin{aligned} \frac{df_0}{dt} + \frac{dg_0}{dt} + \frac{\mu}{\sqrt{2}}\left(\frac{df_1}{dt}\ e^{\lambda_1 x} + \frac{df_2}{dt}\ e^{\lambda_2 x}\right) \\ = \frac{3\mu}{\sqrt{2}}\left(\lambda_1^2\ \frac{df_1}{dt}\ e^{\lambda_1 x} + \lambda_2^2\ \frac{df_2}{dt}\ e^{\lambda_2 x}\right) \\ - k\mu\left(\lambda_1 \frac{df_1}{dt}\ e^{\lambda_1 x} + \lambda_2 \frac{df_2}{dt}\ e^{\lambda_2 x}\right). \end{aligned} \tag{6.65}$$

Since $\lambda_1 \neq \lambda_2$ (due to the assumption that $k > 2\sqrt{m}$), $e^{\lambda_1 x}$ and $e^{\lambda_2 x}$ are linear independent functions of x. Thus, Eq. (6.65) holds for any x if and only if functions $f_0,\ f_1,\ f_2,\ g_0$ give a solution of the system

$$\frac{df_0}{dt} + \frac{dg_0}{dt} = 0\ , \tag{6.66}$$

$$\frac{df_1}{dt} = \left(3\lambda_1^2 - \sqrt{2}\ k\lambda_1\right) f_1\ , \tag{6.67}$$

$$\frac{df_2}{dt} = \left(3\lambda_2^2 - \sqrt{2}\ k\lambda_2\right) f_2\ . \tag{6.68}$$

From Eqs. (6.66–6.68), we obtain:

$$f_0(t) + g_0(t) = C_0\ , \quad f_i(t) = C_i e^{\nu_i t}\ , \quad i = 1, 2 \tag{6.69}$$

where $\nu_i = 3\lambda_i^2 - \sqrt{2}\ k\lambda_i$ and the constants $C_{1,2}$ are determined by the initial conditions.

Note that function $z + w$, as defined by Eqs. (6.61), (6.64) and (6.69), is semi-bounded which agrees with our earlier assumption, cf. the lines below Eq. (6.40). Thus, our analysis has been consistent. Substituting (6.61), (6.64) and (6.69) into Eqs. (6.40), we obtain the following solution of the diffusive predator-prey system (6.8–6.9):

$$u(x,t) = \sqrt{2}\,\frac{\tilde{C}_1\lambda_1\exp(\lambda_1 x + \nu_1 t) + \tilde{C}_2\lambda_2\exp(\lambda_2 x + \nu_2 t)}{1 + \tilde{C}_1\exp(\lambda_1 x + \nu_1 t) + \tilde{C}_2\exp(\lambda_2 x + \nu_2 t)}\ , \qquad (6.70)$$

$$v(x,t) = \sqrt{\frac{2}{\delta}}\,\frac{\tilde{C}_1\lambda_1\exp(\lambda_1 x + \nu_1 t) + \tilde{C}_2\lambda_2\exp(\lambda_2 x + \nu_2 t)}{1 + \tilde{C}_1\exp(\lambda_1 x + \nu_1 t) + \tilde{C}_2\exp(\lambda_2 x + \nu_2 t)} \qquad (6.71)$$

where $\tilde{C}_i = (\mu/\sqrt{2})(C_i/[C_0 + \sigma])$, $i = 1,2$. Thus, relations (6.10–6.11) make the system integrable at the cost of u and v being proportional to each other, $v = u/\sqrt{\delta}$.

It is not difficult to see that, to avoid singularities in the right-hand sides of Eqs. (6.70-6.71), it is necessary that $\tilde{C}_{1,2} > 0$. Introducing new constants as

$$\phi_1 = \frac{1}{\lambda_1}\ln\tilde{C}_1\ , \quad \phi_2 = \frac{1}{\lambda_2}\ln\tilde{C}_2\ , \qquad (6.72)$$

and taking into account that $\bar{u}_{1,2} = \sqrt{2}\lambda_{1,2}$, $\bar{v}_{1,2} = \sqrt{2/\delta}\lambda_{1,2}$ (cf. Eqs. (6.19), (6.22) and (6.63)), the solution (6.70-6.71) takes the following form:

$$u(x,t) = \frac{\bar{u}_1\exp(\lambda_1\xi_1) + \bar{u}_2\exp(\lambda_2\xi_2)}{1 + \exp(\lambda_1\xi_1) + \exp(\lambda_2\xi_2)}\ ,$$

$$v(x,t) = \frac{\bar{v}_1\exp(\lambda_1\xi_1) + \bar{v}_2\exp(\lambda_2\xi_2)}{1 + \exp(\lambda_1\xi_1) + \exp(\lambda_2\xi_2)}$$

where

$$\xi_1 = x - n_1 t + \phi_1\ , \quad \xi_2 = x - n_2 t + \phi_2\ ,$$
$$n_i = -\frac{\nu_i}{\lambda_i} = \sqrt{2}k - 3\lambda_i\ , \quad i = 1,2\ ,$$

and $\lambda_{1,2}$ are given by Eq. (6.63).

In conclusion, it should be mentioned that, since Eqs. (6.8–6.9) are invariant with respect to the transformation $x \to (-x)$, one can expect the existence of the solution symmetrical to (6.28–6.29), i.e., with similar traveling waves but propagating in opposite directions. Indeed, the procedure described above leads to the solution with these properties if we choose minus in both of Eqs. (6.48–6.49). It is readily seen that other options (i.e., plus in (6.48) and minus in (6.49) or vice versa) lead to solutions of (6.8–6.9) which are not nonnegative and, thus, do not seem to have a clear ecological meaning.

6.2 Migration waves in a resource-consumer system

Predator-prey relation is a very important but, of course, not the only possible type of ecological interactions. Another example of exactly solvable model of interacting species is given by a simple resource-consumer system considered by Feltham and Chaplain (2000). In their model, the population of consumer species is assumed to move in space both through isotropic random motion, i.e., diffusion, and with the preferred direction defined by the resource gradient. In its turn, the resource can diffuse and is consumed by the consumer. Aiming to make the model analytically tractable, one has to neglect species multiplication: neither the consumer population nor resource population grow with time. Biologically, it means that the period of time described by this model falls between generations. For simplicity, we also neglect the species' natural mortality, assuming that the mortality rates are low.

The corresponding system of equations is as follows:

$$\frac{\partial n(x,t)}{\partial t} = D_n \frac{\partial^2 n}{\partial x^2} - \frac{\partial}{\partial x}\left(\chi(a) n \frac{\partial a}{\partial x}\right), \tag{6.73}$$

$$\frac{\partial a(x,t)}{\partial t} = D_a \frac{\partial^2 a}{\partial x^2} - k(a) n \tag{6.74}$$

where $n(x,t)$ and $a(x,t)$ give the density of consumer and resource, respectively, at position x and time t. Apparently, due to their meaning, $n(x,t) \geq 0$ and $a(x,t) \geq 0$ for any x and t. D_n and D_a are the diffusion coefficients (assumed to be constant), $k(a)$ is the grazing rate and $\chi(a)$ is the chemotactic response function.

We consider the system dynamics in an infinite space with the following conditions at infinity:

$$a(x \to -\infty, t) = 0, \quad a(x \to \infty, t) = a_0 , \tag{6.75}$$

$$n(x \to \pm\infty, t) = 0 . \tag{6.76}$$

Eqs. (6.73–6.74) must be also supplemented with the initial conditions, i.e., $n(x,0) = g(x)$, $a(x,0) = h(x)$. We do not specify the form of $g(x)$ and $h(x)$ now; however, we assume that they are in agreement with the conditions at infinity (6.75–6.76). Note that conditions (6.76) are consistent with the usual assumption that, at the early stage of the system dynamics, the invasive species is concentrated in a finite domain. Let us mention, however, that model (6.73–6.74) is only partially relevant to biological invasion because it does not take onto account population growth, cf. Section 1.4.

The system (6.73–6.74), although simple enough from the biological point of view, is still difficult to treat analytically, at least, for an arbitrary $k(a)$. In real ecosystems, different species exhibit different types of trophic response.

The most well known are the Holling types I, II and III; however, other types are used as well (see Edwards and Yool (2000) for a brief review). For the purposes of this section, we assume that $k(a) = \mathcal{K} = const$ for $a > 0$ and $k(a) = 0$ for $a = 0$. Biologically, it corresponds to the limiting case of very small values of the half-saturation constants in the Holling types II and III.

The system (6.73–6.74) was studied by Feltham and Chaplain (2000) using perturbation methods. However, in order to make it exactly solvable, we have to assume additionally that the resource is immovable, i.e., $D_a = 0$. This assumption is justified biologically if, for instance, we treat the consumer as a herbivorous species and the resource as a plant species. In this case, the system (6.73–6.74) takes the following form:

$$\frac{\partial n(x,t)}{\partial t} = D_n \frac{\partial^2 n}{\partial x^2} - \frac{\partial}{\partial x}\left(\chi(a) n \frac{\partial a}{\partial x}\right), \tag{6.77}$$

$$\frac{\partial a(x,t)}{\partial t} = -\mathcal{K} n . \tag{6.78}$$

Let us look for a traveling wave solution, i.e., $n(x,t) = \tilde{n}(\xi)$ where $\xi = x - ct$, c being the speed of the wave. The system (6.77–6.78) is then reduced to

$$-c\frac{dn(\xi)}{d\xi} = \frac{d}{d\xi}\left(D_n \frac{dn}{d\xi} - \chi(a) n \frac{da}{d\xi}\right), \tag{6.79}$$

$$c\frac{da(\xi)}{d\xi} = \mathcal{K} n \tag{6.80}$$

(omitting the tildes for notation simplicity).

Integrating (6.79) over space, we obtain

$$-cn = D_n \frac{dn}{d\xi} - \chi(a) n \frac{da}{d\xi} + C_0 \tag{6.81}$$

where C_0 is the integration constant. Here C_0 can be found from conditions at infinity. Indeed, (6.75–6.76) apparently implies, except for some very special cases that are not biologically realistic, that $dn(\xi \to \pm\infty)/d\xi = 0$ and $da(\xi \to \pm\infty)/d\xi = 0$; thus $C_0 = 0$.

Taking n from (6.80) and substituting it into (6.81), we arrive at

$$-c\frac{da}{d\xi} = D_n \frac{d^2 a}{d\xi^2} - \chi(a)\left(\frac{da}{d\xi}\right)^2 . \tag{6.82}$$

Having divided Eq. (6.82) by $da/d\xi$, its integration then leads to the following result:

$$-c\xi = D_n \log\left|\frac{da}{d\xi}\right| - \int \chi(a)\frac{da}{d\xi} d\xi + C_1 \tag{6.83}$$

where C_1 is a new constant and $da/d\xi \neq 0$. Thus, we reduce our further analysis to strictly monotonous profiles of the resource spatial distribution.

From (6.83), we obtain the following equation:

$$\left|\frac{da}{d\xi}\right| = \tilde{C}_1 \exp\left(-\frac{c}{D_n}\,\xi\right) \exp\left(\frac{1}{D_n}\int \chi(a)da\right) . \tag{6.84}$$

Note that the redefined integration constant $\tilde{C}_1 > 0$.

Since, by assumption, $da/d\xi \neq 0$, taking into account (6.75) we obtain that $da/d\xi > 0$ for any ξ. Then $|da/d\xi| = da/d\xi$ and the variables in (6.84) can be separated:

$$\exp\left(-\frac{1}{D_n}\int \chi(a)da\right) da = \tilde{C}_1 \exp\left(-\frac{c}{D_n}\,\xi\right) d\xi . \tag{6.85}$$

The possibility to obtain an exact solution of the system (6.77–6.78) in a closed form thus depends on the form of function $\chi(a)$: the left-hand side of Eq. (6.85) must be integrable! Although the list of appropriate functions is unlikely to be long, one or two examples can be provided easily. Namely, let us consider

$$\chi(a) = \frac{\chi_0}{a + a_1} \tag{6.86}$$

where χ_0 and a_1 are certain constants. A biological justification for this form can be found in Sherratt (1994).

With (6.86), Eq. (6.85) takes the form

$$(a + a_1)^{-\chi_0/D_n}\, da = \tilde{C}_1 \exp\left(-\frac{c}{D_n}\xi\right) d\xi . \tag{6.87}$$

The rest of the analysis depends on the ratio χ_0/D_n.

Special case: $\chi_0/D_n = 1$. From (6.87), we obtain

$$\ln(a + a_1) = \sigma - \frac{D_n\tilde{C}_1}{c} \exp\left(-\frac{c}{D_n}\xi\right) \tag{6.88}$$

(where σ is the second integration constant) and, finally,

$$a(\xi) = \exp\left(\sigma - \exp\left[-\frac{c}{D_n}(\xi - \xi_0)\right]\right) - a_1 \tag{6.89}$$

where $\xi_0 = -(D_n/c)\ln(\tilde{C}_1 D_n/c)$ is the initial "phase" of the wave.

It is readily seen that the boundary conditions (6.75) can only be satisfied if $c > 0$, $\sigma = \ln a_0$ and $a_1 = 0$. Thus, taking into account (6.80), we arrive at the following exact solution:

$$a(\xi) = \exp\left(\log a_0 - \exp\left[-\frac{c}{D_n}(\xi - \xi_0)\right]\right), \tag{6.90}$$

$$n(\xi) = \frac{c^2}{D_n\mathcal{K}} \exp\left[-\frac{c}{D_n}(\xi - \xi_0)\right] \tag{6.91}$$

$$\cdot \exp\left(\log a_0 - \exp\left[-\frac{c}{D_n}(\xi - \xi_0)\right]\right). \tag{6.92}$$

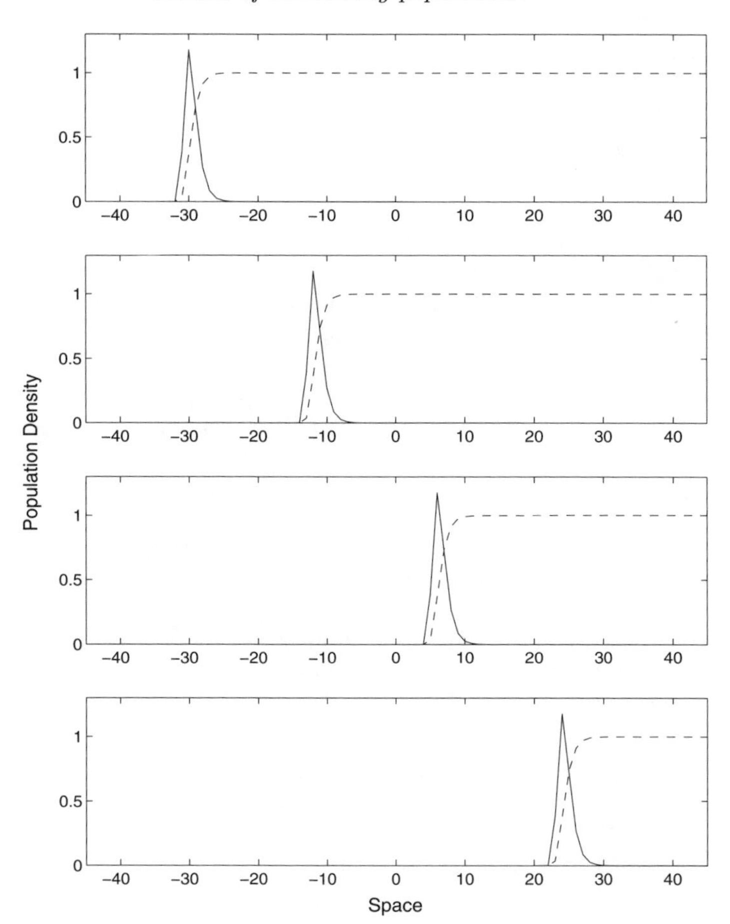

FIGURE 6.6: Traveling wave in a resource-consumer system, cf. (6.77–6.78). Exact solution (6.90–6.91) is shown at equidistant moments, from top to bottom, $t = 0$, $t = 15$, $t = 30$ and $t = 45$. Solid and dashed curves show consumer and resource density, respectively. Parameters are: $D_n = \chi_0 = 1$, $a_0 = 1$, $\mathcal{K} = 0.45$, $c = 1.2$ and $\xi_0 = -30$.

Note that solution (6.90–6.91) does not contain arbitrary constants any more. The constant ξ_0 can be interpreted as the position of the traveling wave at $t = 0$ and thus is defined by the initial conditions.

General case: $\chi_0/D_n \neq 1$. From (6.87), we obtain:

$$\frac{1}{1-(\chi_0/D_n)} \cdot (a + a_1)^{1-(\chi_0/D_n)} = \sigma - \frac{D_n \tilde{C}_1}{c} \exp\left(-\frac{c}{D_n}\,\xi\right) \quad (6.93)$$

so that the expression for a is

$$a(\xi) = \left(\tilde{\sigma} - \frac{D_n\tilde{C}_1}{c}\left(1 - \frac{\chi_0}{D_n}\right)\exp\left(-\frac{c}{D_n}\,\xi\right)\right)^{1/(1-\chi_0/D_n)} \tag{6.94}$$
$$- a_1$$

where $\tilde{\sigma} = (1 - (\chi_0/D_n))\sigma$ is a new constant.

Note that, for biological reasons, we are interested only in bounded solutions. Apparently, solution (6.94) is bounded if and only if the exponent in the right-hand side is negative, i.e., for $\chi_0 > D_n$. Taking that into account and introducing the initial phase of the wave as

$$\tilde{\xi}_0 = -\frac{D_n}{c}\ln\left[\frac{D_n\tilde{C}_1}{c}\left(\frac{\chi_0}{D_n} - 1\right)\right],$$

from (6.94), we obtain:

$$a(\xi) = \left(\tilde{\sigma} + \exp\left[-\frac{c}{D_n}(\xi - \tilde{\xi}_0)\right]\right)^{1/(1-\chi_0/D_n)} - a_1\,. \tag{6.95}$$

Taking into account the conditions at infinity and the relation (6.80), which results in $a_1 = 0$, from (6.94) we finally obtain another exact traveling wave solution of the system (6.77–6.78):

$$a(\xi) = \left(\tilde{\sigma} + \exp\left[-\frac{c}{D_n}(\xi - \tilde{\xi}_0)\right]\right)^{D_n/(D_n-\chi_0)}, \tag{6.96}$$

$$n(\xi) = A\exp\left[-\frac{c}{D_n}(\xi - \tilde{\xi}_0)\right] \tag{6.97}$$

$$\cdot\left(\tilde{\sigma} + \exp\left[-\frac{c}{D_n}(\xi - \tilde{\xi}_0)\right]\right)^{(\chi_0)/(D_n-\chi_0)} \tag{6.98}$$

where

$$A = \frac{c^2}{D_n\mathcal{K}}\left(\frac{D_n}{\chi_0 - D_n}\right) \tag{6.99}$$

is a coefficient,

$$\tilde{\sigma} = a_0^{1-(\chi_0/D_n)} \tag{6.100}$$

and $c > 0$.

Fig. 6.6 shows the snapshots of the exact solution (6.90–6.91) obtained at equidistant moments with the time-step $\Delta t = 15$ for a hypothetical parameter set $c = 1.2$, $a_0 = 1$, $\mathcal{K} = 0.45$, $D_n = \chi_0 = 1$ and $\xi_0 = -30$. Thus, the solution has an apparent biological meaning: it describes species migration along the resource gradient. It is readily seen that the solution (6.96–6.97) possesses similar properties; for the sake of brevity, we do not show it here.

Chapter 7

Some alternative and complementary approaches

In the previous chapters, we have revisited a few methods that can be used to obtain exact solutions of nonlinear diffusion-reaction equations, and gave several examples of exactly solvable models. Some of the methods are rather general, e.g., the method of piecewise-linear approximation, and their practical application is only limited by the amount of tedious calculations to be made; the others are ad hoc methods that are likely to be successful when applied to a few particular cases of high biological relevance. As a whole, this set of approaches and relevant examples provides, when applied properly, an efficient tool for studying biological invasions.

Meanwhile, it must be mentioned that exactly solvable models give only a relatively small part of all nonlinear models that arise in the studies of species spread. A general observation regarding mathematical modeling is that the less information a given method yields the wider is the class of problems it can be applied to. Clearly, an exact solution gives exhaustive information about the problem under study. In case the expectations are restricted to a particular solution's property, the class of analytically treatable models can be much wider.

Regarding biological invasion, the quantity of primary importance is the rate of species spread which, in the case of a traveling front propagation, coincides with the front speed. The speed of the waves is important as well in other traditional applications of diffusion-reaction models such as combustion and flame propagation (Zeldovich and Barenblatt, 1959; Zeldovich et al., 1980; Volpert et al., 1994), chemical engineering (Aris, 1975) and neurophysiology (Scott, 1977); for a more general scope and a wider list of references see also Britton (1986) and Grindrod (1996). For these reasons, considerable work has been done and extensive literature has been published concerning the wave speed in diffusion-reaction systems and the methods of its calculation.

Another problem concerns the traveling wave solutions themselves. Apparently, reduction of PDE to ODE leaves aside a great majority of relevant solutions of the original problem that do not possess the required symmetry. However, a remarkable point is that traveling wave solutions often act as "attractors" for initial conditions from a wide class. That brings forward the issue of convergence to relevant traveling wave solution. Interestingly, it appears to be tightly related to the speed of the wave propagation.

As a whole, mathematics of biological invasion and species spread is a vast scientific field. On the one side, it is rooted in a more general theory of traveling waves in diffusion-reaction systems and, eventually, in the theory of nonlinear partial differential equations. On the other side, it involves a lot of knowledge from biology, ecology and environmental science which greatly affect the background of the modeling approach, cf. Chapter 2. Thus, it would be impossible to give a review of all relevant issues here. Instead, the goal of this chapter is much more modest. Namely, we are going to briefly outline a somewhat more general mathematical framework for the actual content of our book, i.e., exactly solvable models. In particular, we consider what kind of information about the species spread can be obtained in non-integrable cases. Also, we will show how possible application of exactly solvable models can sometimes be extended beyond their formal capacities.

7.1 Wave speed and the eigenvalue problem

In most parts of this book, our analysis has been based on the following single-species model of population dynamics:

$$u_t(x,t) = u_{xx} + F(u) \tag{7.1}$$

(in dimensionless variables). Due to the phenomenon of convergence, which we address separately in the next section, application of Eq. (7.1) to biological invasion can in many cases be reduced to consideration of its traveling wave solutions, i.e., solutions of the equation

$$\frac{d^2U}{d\xi^2} + c\frac{dU}{d\xi} + F(U) = 0 \tag{7.2}$$

where $U = U(\xi)$, $\xi = x - ct$ and c is the wave speed.

Equation (7.2) describes the geographical spread of the invasive species provided the conditions at infinity correspond to the equilibrium states of the system, i.e., to the zeros of function F. For biological reasons, $F(U)$ must allow for at least two equilibrium states, $F(0) = F(1) = 0$. Then, assuming here without any loss of generality that the alien species spreads along axis x, the conditions are

$$U(\xi \to -\infty) \;=\; 1, \quad U(\xi \to \infty) \;=\; 0. \tag{7.3}$$

In spite of the fact that Eq. (7.2) may at first glance look rather simple, it is not possible to obtain a nontrivial closed-form solution of Eq. (7.2) for an arbitrary $F(U)$. However, even when an explicit solution is not available, it is sometimes possible to obtain the wave speed.

Equation (7.2) with (7.3) can be regarded as a nonlinear eigenvalue problem: find all values of c for which Eq. (7.2) has a nonnegative solution satisfying conditions (7.3). There are a few cases when this problem can be solved rigorously. In particular, it was shown by Kolmogorov et al. (1937) in their seminal paper that, for F described by conditions

$$F(u) > 0 \ \text{ for } \ 0 < u < 1, \quad F(u) < 0 \ \text{ for } \ u > 1, \tag{7.4}$$

$$F'(0) = 1 \ > \ 0, \quad F'(u) \ < \ 1 \ \text{ for } \ u > 0, \tag{7.5}$$

cf. "generalized logistic growth," the eigenvalue problem has a continuous spectrum: for any $c \geq 2$ there exists a solution with necessary properties.

For a function F with other features, the spectrum can be essentially different. For instance, in case $F(u)$ possesses an intermediate zero, i.e., $F(\beta) = 0, \ 0 < \beta < 1$, so that

$$F(u) < 0 \quad \text{for} \quad 0 < u < \beta \ \text{ and } \ u > 1 \ , \tag{7.6}$$

$$F(u) > 0 \quad \text{for} \quad \beta < u < 1 \tag{7.7}$$

(which corresponds to the strong Allee effect), the spectrum consists of a single value c_0. This result was originally obtained by Zeldovich (1948) in connection to the flame propagation problem and was later generalized by Aronson and Weinberger (1975, 1978).

The full consideration of the problem requires analysis of the global behavior of the trajectories in the corresponding phase plane. However, one essential difference between these two cases can be conceived by means of a simple linear analysis. Let us consider the solution of Eq. (7.2) far in front of the front, i.e., where $U \ll 1$. Then the growth function can be linearized, $F(U) \approx F'(0)U$, and Eq. (7.2) takes the form

$$\frac{d^2U}{d\xi^2} + c\frac{dU}{d\xi} + F'(0)U = 0 \tag{7.8}$$

assuming that $F'(0) \neq 0$.

Equation (7.8) is linear; therefore, its general solution is either

$$U(\xi) = C_1 e^{\lambda_1 \xi} + C_2 e^{\lambda_2 \xi} \tag{7.9}$$

(if $\lambda_1 \neq \lambda_2$) or

$$U(\xi) = (C_1 + C_2\xi)\, e^{\lambda_1 \xi} \tag{7.10}$$

(if $\lambda_1 = \lambda_2$) where $C_{1,2}$ are arbitrary constants and $\lambda_{1,2}$ are the solutions of the following equation:

$$\lambda^2 + c\lambda + F'(0) = 0 \tag{7.11}$$

so that

$$\lambda_{1,2} = \frac{1}{2}\left(-c \pm \sqrt{c^2 - 4F'(0)}\right). \tag{7.12}$$

Since U is the population density, it cannot be negative; therefore, the solution cannot oscillate around $U = 0$. Correspondingly, it means that $\lambda_{1,2}$ must be real which is only possible for

$$c^2 \geq 4F'(0). \tag{7.13}$$

In case of the generalized logistic growth, from (7.13) we immediately obtain the minimum possible value of the wave speed:

$$c_{min} = 2. \tag{7.14}$$

Note that conditions (7.13–7.14) obtained in this way neither guarantee that the solution of problem (7.2–7.3) exists for any $c \geq 2$ nor that $c = 2$ is the exact lower bound of the spectrum; the only conclusion that can actually be made is that there can be no traveling wave solution for $c < 2$. Nevertheless, Eq. (7.14) appears to coincide with the results of a more comprehensive analysis. Thus, the wave speed is determined by the profile properties at the leading edge, i.e., far in front of the front. For that reason, traveling waves in the population described by (7.2) with (7.4–7.5) are sometimes referred to as the "pulled" waves, cf. Hadeler and Rothe (1975).

In case of the Allee effect, however, $F'(0) < 0$ and inequality (7.13) only leads to a trivial conclusion that $c^2 \geq 0$. The fact that linear analysis in vicinity of $U = 0$ does not bring any information indicates that the wave speed is determined not by the leading edge behavior but by the properties of the wave profile inside the transition region. A similar inference can be also made from consideration of wave blocking conditions, cf. (2.22) and (2.23). The traveling waves of this type are called "pushed" waves.

Note that condition (7.13) does not actually say anything about the sign of c, i.e., about the direction of wave propagation. It is readily seen from Eq. (7.2) that the sign of c is fully determined by the properties of the growth function F. Indeed, multiplying (7.2) by $dU/d\xi$, integrating over space and making use of (7.3), we obtain:

$$c\int_{-\infty}^{\infty}\left(\frac{dU}{d\xi}\right)^2 d\xi = \int_0^1 F(U)dU\ . \tag{7.15}$$

Obviously, the sign of c coincides with the sign of the integral in the right-hand side so that it can be different only in case F is an alternating-sign function. As an immediate consequence, this means that the wave always propagates towards the domain with low population density in case of the generalized logistic growth but can propagate in either direction in case of the strong Allee effect.

The above results relate to the single-species model with linear diffusion. Some generalization appears possible to the case of density-dependent diffusivity. In particular, for $D(U) = U$ and $F(U) = U(1-U)$, cf. Eq. (5.6), Sánchez-Garduño and Maini (1994) showed that the corresponding eigenvalue problem has a continuous spectrum as well, although the lower bound of the spectrum is different. The traveling wave solution exists only for $c \geq c_{min} = 1/\sqrt{2}$ so that $c = 1/\sqrt{2}$ corresponds to the sharp front and $c > 1/\sqrt{2}$ corresponds to the smooth fronts.

7.2 Convergence of the initial conditions

In a rigorous mathematical sense, the traveling wave solutions of diffusion-reaction equation (7.1) arise for particular initial conditions, i.e., the conditions describing the wave profile at the moment $t = 0$. In case of integrable models, these conditions are immediately obtained by means of letting $t = 0$ in the exact solution expression. Remarkably, however, the meaning of the traveling wave solution is not exhausted by this rather special case. There are numerous results, cf. Volpert et al. (1994), showing that initial conditions from a wide class exhibit convergence to a relevant traveling wave solution in the large-time asymptotics.

From the standpoint of species invasion or colonization, the most biologically reasonable initial conditions to Eq. (7.1) are those that are described by functions of either finite or semi-finite support. One of the most comprehensive mathematical studies of population dynamics initiated by conditions of that type was done by Kolmogorov et al. (1937). In particular, they considered the initial condition of "transitional" type:

$$u(x,0) = 1 \text{ for } x \leq x_0, \quad u(x,0) = 0 \text{ for } x \geq x_0 + \Delta, \tag{7.16}$$

$$u(x,0) = \phi(x) \geq 0 \text{ for } x_0 < x < x_0 + \Delta \tag{7.17}$$

where $\phi(x)$ is a certain function and Δ is a parameter giving the width of the transition region. Kolmogorov et al. (1937) showed that, for a population with the local growth described as (7.4–7.5), conditions (7.16–7.17) always converge to a traveling wave propagating with the minimum possible speed $c = c_{min} = 2$.

The situation stays essentially the same in case of initial conditions of finite support,

$$u(x,0) = 0 \text{ for } x \leq x_0 \text{ and } x \geq x_0 + \Delta, \tag{7.18}$$

$$u(x,0) = \phi_1(x) \geq 0 \text{ for } x_0 < x < x_0 + \Delta, \tag{7.19}$$

up to the apparent difference that evolution of (7.18–7.19) leads to formation of two traveling fronts propagating in opposite directions with speed $c = \pm 2$, respectively.

We should recall here that the nonlinear eigenvalue problem (7.2–7.3) with $F(u)$ describing the generalized logistic growth has a solution for any $c \geq 2$. The question thus remains about the waves propagating with a speed larger than c_{min}, i.e., what kind of initial conditions they may correspond to. Indeed, it appears that the value of speed that is actually "chosen" by the traveling population wave is determined by certain properties of the initial species distribution. Specifically, it was shown by Rothe (1978), see also McKean (1975) and Larson (1978), that the speed depends on its large-distance asymptotical behavior: assuming that it exhibits an exponential decay, i.e., $u(x,0) \sim e^{-sx}$ for $x \to +\infty$, the wave speed is larger the smaller is s:

$$c = 2 \text{ if } s \geq 1 \text{ and } c = s + \frac{1}{s} \text{ if } s < 1. \tag{7.20}$$

Note that, although initial conditions defined in an infinite domain are by themselves not of much relevance to biological invasion, relation (7.20) is very useful in the sense that it helps to understand what kind of function can be possibly used to describe $u(x,0)$ for modeling purposes. In particular, it is immediately seen that convergence to the wave propagating with the minimum speed takes place when the initial conditions are described by the Gaussian distribution, cf. Section 4.2.

The system dynamics with respect to initial conditions changes significantly when the population growth is damped by the strong Allee effect so that $F(u)$ is not positively defined in interval $(0,1)$ any more, cf. (7.6–7.7). In this case, the eigenvalue problem (7.2–7.3) has the only solution corresponding to a certain $c = c_0$. As a result, not only for the transitional initial conditions (7.16–7.17), but also for a more general type,

$$u(x,0) \to 1 \text{ for } x \to -\infty \text{ and } u(x,0) \to 0 \text{ for } x \to \infty, \tag{7.21}$$

the solution of Eq. (7.1) always converges to a traveling wave propagating with speed c_0. Note that the sign of c_0 can now be different depending on the sign of $\int_0^1 F(u)du$; see the previous section.

In case of finite initial conditions, there can be two different dynamical regimes depending on Δ and $\Phi_1 = \max \phi_1(x)$ (cf. the problem of critical aggregation, Section 4.3). This problem was first considered rigorously by Kanel (1964). For a specific case $\phi_1(x) \equiv 1$, he showed that, provided $\int_0^1 F(u)du > 0$, the initial distribution (7.18–7.19) converges to two traveling waves propagating in opposite directions in case Δ is sufficiently large but it approaches zero in case Δ is sufficiently small.

We want to emphasize here that, in spite of considerable progress in mathematics of diffusion-reaction systems (e.g., see Volpert et al. 1994), practically useful algorithms helping to calculate the critical width and magnitude of the initial distribution in a more or less general case are lacking and the problem of distinguishing between the two regimes in terms of the corresponding initial conditions is largely open.

In conclusion to this section, we want to mention that, although the *rate* of convergence is mainly determined by the properties of function F and, as such, can be studied under rather general assumptions, cf. Larson (1978), the *characteristic time* of convergence τ_c significantly depends on the particulars of initial conditions and thus can vary greatly from case to case. A general tendency is that this time is larger for the populations with the Allee effect than for the populations with the generalized logistic growth; in particular, $\tau_c \to \infty$ for $\Delta \to \Delta_{cr}$. As a result, in a spatially-bounded system it may happen that the time required by wave formation is on the same order or larger than the time of wave propagation through the domain (Ognev et al., 1995).

7.3 Convergence and the paradox of linearization

In order to get a deeper insight into convergence of initial conditions to a traveling wave, in this section we consider an illustrative example that we call the "paradox of linearization." As above, we consider a single-species model of population dynamics, cf. (7.1), with $F(u)$ corresponding to the generalized logistic growth. For the purposes of this section, it is more instructive to consider the problem in the original (dimensional) variables so that $F'(0) = \alpha > 0$. We assume that the initial species distribution is described by a function of finite support defined in the domain situated around $x = 0$. We restrict our consideration to population spreading along the axis x; its spread against the axis x takes place in the same manner up to the change $x \to -x$.

Clearly, at a position in space sufficiently far away from the place of the initial distribution, the population density u is small and the equation can be linearized:

$$u_t(x,t) = Du_{xx} + \alpha u \ . \tag{7.22}$$

It is readily seen that its solution is

$$u(x,t) \simeq \frac{1}{\sqrt{t}} \exp\left(-\frac{x^2}{4Dt} + \alpha t\right) \tag{7.23}$$

(cf. Chapter 9) where we have omitted the constant coefficient for convenience.

On the other hand, in the large-time limit, finite initial conditions are known to converge to a traveling wave. In front of the front, the density $u(x,t) = U(\xi)$ is small so that $F(U) \approx \alpha U$ and the dynamics is again described by a linear differential equation but of essentially different type:

$$D\frac{d^2U}{d\xi^2} + c\frac{dU}{d\xi} + \alpha U = 0 \tag{7.24}$$

where $\xi = x - ct$. In case of a wave propagating with minimum speed (as it always happens for finite initial conditions), the solution of Eq. (7.24) has the following large-distance asymptotics:

$$u(x,t) = U(\xi) \;\simeq\; \xi \exp\left(-\frac{c_{min}}{2D}\,\xi\right) \tag{7.25}$$

omitting the constant coefficient and taking into account that $c_{min} = 2\sqrt{D\alpha}$, cf. (7.10) and (7.12).

Now, we arrive at a paradox: for any fixed moment t, expressions (7.23) and (7.25) give apparently different rates of solution decay, i.e., on the order of $\exp(-x)$ for the traveling wave solution and on the order of $\exp(-x^2)$ for the original diffusion problem. We want to emphasize that, while (7.24) and (7.25) are only relevant when t is sufficiently large, i.e., after the wave has formed, application of Eq. (7.22) is not necessarily restricted to early stages of the system dynamics and (7.23) is valid for any t provided the distance x is large enough.

In order to resolve this seeming contradiction, we transform (7.23) as follows:

$$\begin{aligned} u(x,t) &\simeq \exp\left(-\frac{x^2}{4Dt} + \alpha t - \frac{1}{2}\ln t\right) \\ &= \exp\left(-\frac{1}{4Dt}\,[x - \tilde{c}t][x + \tilde{c}t]\right) \end{aligned} \tag{7.26}$$

where

$$\begin{aligned} \tilde{c} \;&=\; 2\sqrt{D\alpha}\left[1 - \left(\frac{1}{2\alpha}\right)\frac{\ln t}{t}\right]^{1/2} \\ &=\; c_{min}\left[1 - \left(\frac{1}{2\alpha}\right)\frac{\ln t}{t}\right]^{1/2} . \end{aligned} \tag{7.27}$$

Note that $\tilde{c}(t) \to c_{min}$ for $t \to \infty$.

We then introduce a new variable, $\tilde{\xi} = x - \tilde{c}t$. Correspondingly, Eq. (7.26) takes the form

$$u(x,t) \;=\; \tilde{U}(\tilde{\xi},t) \simeq \exp\left(-\frac{\tilde{c}}{2D}\,\tilde{\xi} - \frac{\tilde{\xi}^2}{4Dt}\right). \tag{7.28}$$

Expression (7.28) includes both of the asymptotics (7.23) and (7.25). Which of them actually takes place depends on $\tilde{\xi}$ and t. Namely, Eq. (7.28) describes

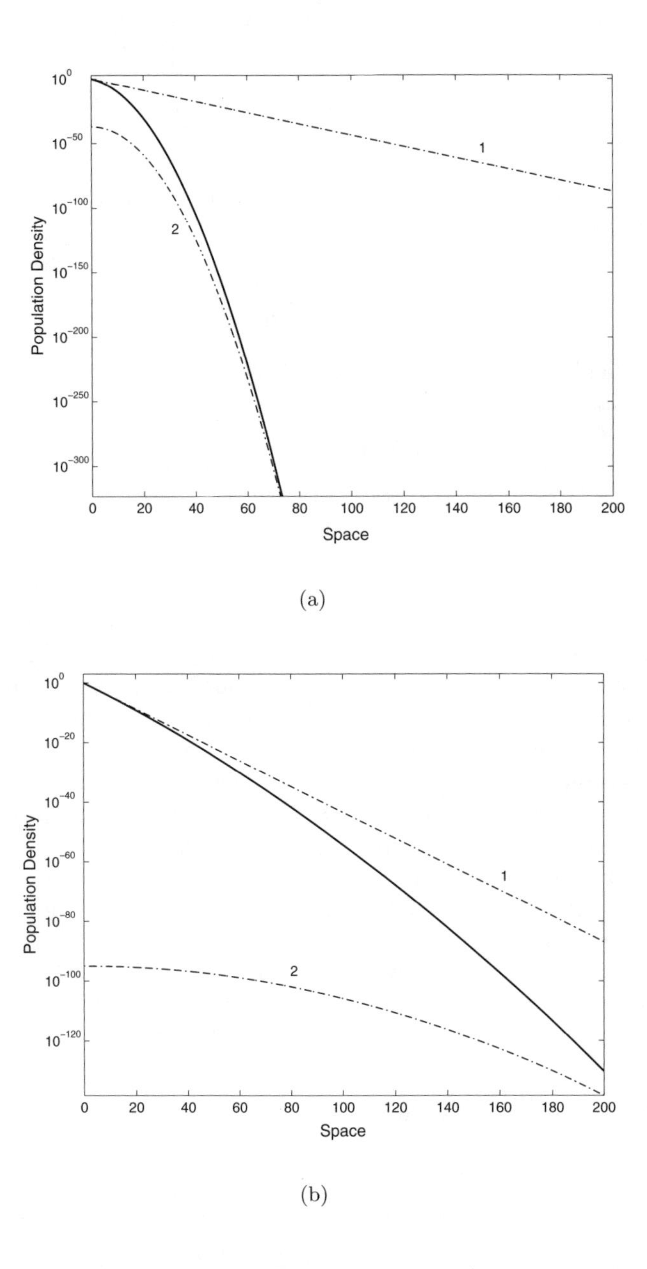

FIGURE 7.1: Population density versus space (semilogarithmic plot) in a traveling population front as given by (7.28) (solid curve) for (a) $t = 2$ and (b) $t = 100$; curves 1 and 2 show exponential and Gaussian asymptotics, respectively. Apparently, in the course of time the domain where Gaussian asymptotics is applicable shrinks toward the leading edge.

the exponential decay in case the second term in the parentheses is negligible with respect to the first term,

$$\frac{\tilde{c}}{2D}\,\tilde{\xi} \gg \frac{\tilde{\xi}^2}{4Dt}\,, \tag{7.29}$$

e.g., for sufficiently large t, and it describes the rate of decay as $\exp(-x^2)$ in the opposite case,

$$\frac{\tilde{c}}{2D}\,\tilde{\xi} \ll \frac{\tilde{\xi}^2}{4Dt}\,, \tag{7.30}$$

e.g., for sufficiently large $\tilde{\xi}$. The change between the two different asymptotics thus takes place for

$$\frac{\tilde{c}}{2D}\,\tilde{\xi} \simeq \frac{\tilde{\xi}^2}{4Dt}\,, \tag{7.31}$$

i.e., for

$$\tilde{\xi}_* \simeq 2\tilde{c}t \tag{7.32}$$

so that on the left of $\tilde{\xi}_*$ the asymptotics is exponential (provided that the density is small enough and the linear approximation is valid) while on the right of $\tilde{\xi}_*$ it is of the Gaussian type; see Fig. 7.1. We want to emphasize that, due to the nature of relations (7.29) and (7.30), $\tilde{\xi}_*$ gives not the exact position but rather the order of magnitude where this change takes place.

Obviously, the "critical coordinate" $\tilde{\xi}_*$ moves along the wave profile toward the leading edge so that the domain with the traveling wave-type asymptotics grows with time. Expression (7.28) thus describes the approach of the initial conditions to the traveling wave solution. Interestingly, it predicts a different type of convergence for the wave speed and the wave profile: while the approach to the speed takes place with the same rate $(1/t)\ln t$ for both types of the asymptotics, convergence to the wave profile takes place inhomogeneously in space. Remarkably, the results of our heuristic analysis appear to be in very good agreement with the results of a more rigorous analysis, cf. McKean (1975) and Larson (1978).

7.4 Application of the comparison principle

Exact solutions of nonlinear partial differential equations are typically obtained either for a specific form of nonlinearity or for particular initial conditions. In more applied biological or ecological studies, this circumstance is often mistaken for a reason to underestimate the meaning of exactly solvable models. An alternative theoretical tool is usually seen in extensive numerical simulations; however, with all proper respect to potency of computer experiment, simulations alone can hardly be an adequate substitute to a rigorous mathematical study.

As a matter of fact, this foible of exact solutions can be turned to their strength by means of the comparison principle. The formulation of this principle can be slightly different; here we mainly stick to the one by Volpert and Khudyaev (1985). A detailed consideration of related issues can be found in Protter and Weinberger (1984).

Theorem 7.1. *Let $u_1(\mathbf{r},t)$ and $u_2(\mathbf{r},t)$ are nonnegative solutions of the equations*

$$\frac{\partial u_1(\mathbf{r},t)}{\partial t} = D\nabla^2 u_1 + F_1(\mathbf{r},t,u_1) \ , \tag{7.33}$$

$$\frac{\partial u_2(\mathbf{r},t)}{\partial t} = D\nabla^2 u_2 + F_2(\mathbf{r},t,u_2) \tag{7.34}$$

for $\mathbf{r} \in \mathbf{R}^n$ and $t > 0$, with the initial conditions

$$u_1(\mathbf{r},0) \ = \ \phi_1(\mathbf{r}), \quad u_2(\mathbf{r},0) \ = \ \phi_2(\mathbf{r}), \tag{7.35}$$

and the conditions at infinity as

$$u_{1,2}(\mathbf{r},t) \ \le \ M \ < \ \infty \quad for \ \ |\mathbf{r}| \to \infty \ , \tag{7.36}$$

where M is a certain constant.

If the following conditions are satisfied:

$$\phi_1(\mathbf{r}) \ \le \ \phi_2(\mathbf{r}), \quad F_1(\mathbf{r},t,u) \ \le \ F_2(\mathbf{r},t,u) \ , \tag{7.37}$$

then $u_1(\mathbf{r},t) \le u_2(\mathbf{r},t)$ for all $\mathbf{r} \in \mathbf{R}^n$ and $t > 0$.

The proof can be found in the above-mentioned book; for the sake of brevity, we do not reproduce it here.

Now, how can Theorem 7.1 be used to facilitate application of exact solutions to study biological invasion and species spread in a general case? Let F_1 be a realistic parameterization of the local growth rate of a given invasive species, e.g., reconstructed from available biological data, and F_2 is one of the functions that makes the model exactly solvable. We assume that $F_1 \le F_2$ for all values of their arguments; see Fig. 7.2. In most cases, exactly solvable models describe propagation of traveling waves in 1-D space so an appropriate form of Eq. (7.34) is its one-dimensional reduction. In a more realistic 2-D case, it corresponds to propagation of a plane wave along or against axis x. The initial distribution of the invasive species is usually described by a function of compact support defined in a certain domain inside $\mathbf{R}^2$. Apparently, such function can always be majorized by the two-dimensional extension of the 1-D initial conditions corresponding to the traveling wave solution. Exact

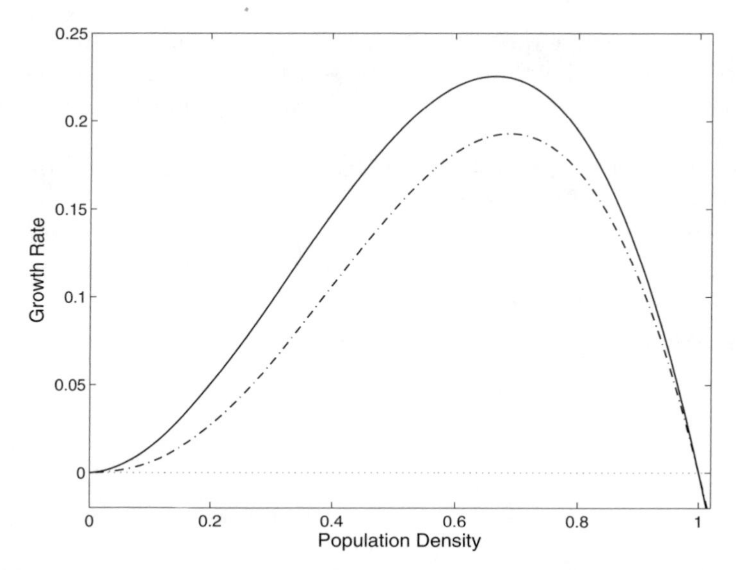

FIGURE 7.2: An example of relation between a "real" growth function $F_1(u)$ (dashed-and-dotted curve) and a function $F_2(u)$ that makes the model exactly solvable; details are given in the text.

solution $u_1(x,t) = U(x - ct)$ then gives an upper bound of the population density of the invasive species and speed c gives an upper bound of the invasion rate. Similar arguments can be applied to other types of exact solution as well, cf. Sections 4.2 and 4.3.

Note that, due to generality of the theorem conditions, the approach described above works as well when F_1 depends explicitly on space and/or time, e.g., as a result of environmental heterogeneity and seasonal or climatic changes. However, the accuracy of the invasion rate estimate based on a relevant exact solution is likely to be much lower in that case.

Chapter 8

Ecological examples and applications

There are many examples of successful application of rigorous mathematical results to understanding and prediction of invasive species spread, e.g., see Skellam (1951), Lubina and Levin (1988), Okubo et al. (1989), Andow et al. (1990), Lewis and Kareiva (1993), Shigesada et al. (1995), Owen and Lewis (2001), and also Shigesada and Kawasaki (1997) and the references therein. In most cases, comparison between theory and data is based, directly or indirectly, on the equation for the minimum speed of the population front of the invasive species with logistic growth, i.e., $c = 2\sqrt{D\alpha}$ (see Sections 2.1 and 7.1). Apparently, application of diffusion-reaction equations to biological invasion is not exhausted by this relatively simple case. A question of particular importance is what factors can possibly modify the front speed: either speeding up or slowing down, or even blocking/reversing species invasion. A few such factors have been addressed above by means of exactly solvable models; see Sections 4.1, 4.2 and 6.1. In this chapter, we are going to confront the mathematical results with some relevant ecological data in order to verify the theoretical predictions and to further check the models' capacities.

It must be mentioned that any instructive comparison between theory and practice is only possible when both of them use the same quantities to describe the phenomenon under study. In case of biological invasion, the main quantity used in field ecology is the range of the alien species while the main quantity used in theoretical approaches is the radius R of invaded area. The relation between these values is not always straightforward. First, most of the analytically solvable models deal with one-dimensional systems while in nature species spread usually takes place in two dimensions [but see Lubina and Levin (1988) for an example of 1-D biological invasion]. Although the radially expanding front is expected to converge to the plane wave in the large-t or large-R limit, the practical question concerning what R can be regarded as large often brings a degree of uncertainty. Second, from the standpoint of real ecosystems the definition of the invaded area radius itself can be ambiguous as well. The matter is that, due to the impact of various factors, e.g., environmental heterogeneity, the border of the species range is rarely close to a circle. Thus, calculation of the radius should either involve averaging over different directions of species spread or using an alternative definition of the area radius. This problem was addressed by several authors, e.g., see Skellam (1951), Andow et al. (1990) and Shigesada and Kawasaki (1997).

There are a lot of cases of species invasion where field data either are far too incomplete or available in a form not suitable for immediate comparison with theoretical predictions. Although working on these cases is a challenging problem, it clearly lies beyond the scope of this book. Since here we are primarily concerned with exactly solvable models, for comparison between theory and data we have selected a few examples where the uncertainties mentioned above were solved successfully. (Note that we are not going into details of data analysis because it is out of the scope of this book; those who are interested can find them in original publications cited in the text.) We want to emphasize, however, that applicability and usefulness of the exactly solvable models clearly reach far beyond these cases and more examples can be given by more focused studies.

8.1 Invasion of Japanese beetle in the United States

Probably one of the most famous cases of biological invasion has been the invasion of Japanese beetle in the United States in the first half of the twentieth century. Although the exact date of its introduction is unknown, as it is often happens with invasion of alien species, it is thought to have been brought into the US around 1911 from Japan with some commercial plant species. Having started its spread from a farm in New Jersey, this species increased its range to about 50,000 km^2 in less than thirty years and eventually spread over the whole eastern United States. The dynamics of invasion during the first several years is shown in Fig. 8.1 and the population density inside the infested area promptly reached very large values. The spread of this pest resulted in a virtual wipeout of many plant species ranging from clover to apple-tree, and the economic losses were tremendous. More details can be found in Elton (1958).

Apart from the fact that the Japanese beetle invasion caused a severe damage to agriculture and that it is well documented (which makes this case suitable for analysis), there is a feature that makes this case especially interesting for our purposes. Namely, although it may not be immediately seen from Fig. 8.1, a closer inspection of the data shows that, during the first decade of the invasion, the radius of the infested area was increasing with accelerating speed; see Fig. 8.2 where asterisks show the averaged radius of the invaded area versus time as it was obtained in field observations. As it has already been mentioned above, the "standard" diffusion-reaction models seem to predict a constant-rate spread and, as such, might be thought not capable to account for this phenomenon. In this section, we will show that this is not true and that front propagation at increasing speed is an intrinsic property of diffusion-reaction models – at least at a certain stage of the species spread.

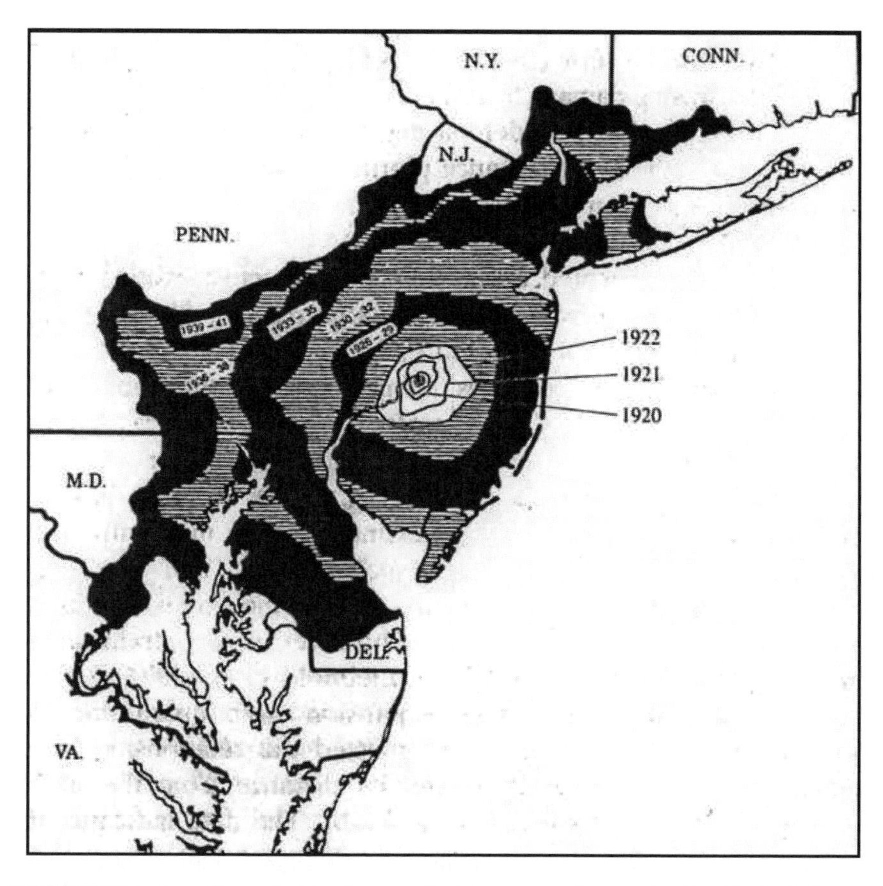

FIGURE 8.1: Geographical spread of Japanese beetle in the United States from the place of its original introduction. Alternating grey and black colors show the areas invaded during successive time intervals (from United States Bureau of Entomology and Plant Quarantine, 1941).

An attempt to explain the nature of accelerating waves has been made by means of linking them to "non-Gaussian" diffusion when every single event of dispersal follows not the normal distribution but a distribution with much slower rate of decay at large distances (see Section 2.2). Mathematically, it means that the models should be based on integral-difference equations, not diffusion-reaction ones. Indeed, for some plant species this approach works very well, cf. Clark et al. (1998). For insect species, however, the situation is different. While in the case of seed or pollen spreading its fat-tailed dispersal can be linked to the peculiarity of turbulent wind mixing, in the case of insects, they are not just passively born by the wind. A mechanistic theory of insects' dispersal accounting for their ability to self-motion is lacking and the existing data are usually of poor accuracy so that they can be easily fitted by

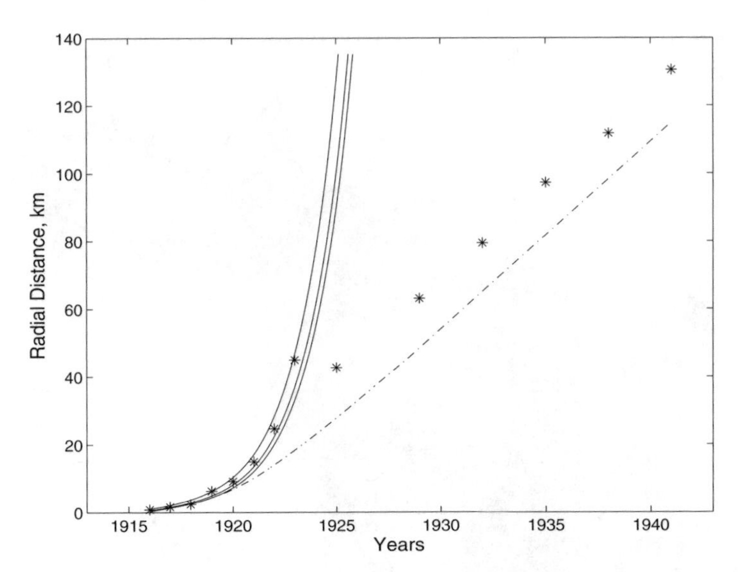

FIGURE 8.2: Comparison between the exact solution (4.40), (4.46) and (4.47) and the data on Japanese beetle range expansion. Asterisks show averaged radius of the invaded area versus time as it was obtained in field observations. The solid lines show $R(t)$ as given by (4.46) for different values of parameter R_{in}, from left to right: $R_{in} = 2.0$, $R_{in} = 1.0$, $R_{in} = 0.1$. The dashed-and-dotted line shows the results of numerical integration of Eq. (4.33) with $f(u) = f_\gamma(u)$ for $\gamma = 0.033$ and the same values of R_{in} (with permission from Petrovskii and Shigesada, 2001).

virtually any distribution (Kot et al., 1996). Moreover, in Section 2.2 we have shown that the species spread at a speed higher than $c = 2\sqrt{D\alpha}$ (which has been observed for some insect species and is sometimes considered as a sign of non-Gaussian diffusion) can still be described by diffusion-reaction equations.

Another explanation of biological invasion with increasing rates of species range expansion can be provided if we assume that, in some cases, this regime corresponds to an early stage of the invasion, before the stationary traveling wave is formed. In Section 4.2.1 we showed (see Fig. 4.10) that, indeed, at the beginning of invasion the population spreading can take place with increasing speed. Moreover, when considering a population invasion in a bounded domain, for certain parameter values it may happen that the constant-rate stage is never reached.

It should be mentioned that one simple mechanism of species spread at increasing speed is readily identified as soon as we take into account that the spread of the invasive species normally takes place in two dimensions. The speed c_{cyl} of the front propagation in the case of cylindrical symmetry is related to the speed c of the plane wave propagation by means of the following equation:

$$c_{cyl} = c - \frac{D}{R} \tag{8.1}$$

(cf. Mikhailov, 1990) where $R = R(t)$ is the radius of the invaded area. Since $c_{cyl} = dR/dt$, Eq. (8.1) is easily solved giving the area radius dependence on time in an implicit way:

$$t = \frac{1}{c}\left[(R - R_{in}) + \frac{D}{c}\ln\left(\frac{cR - D}{cR_{in} - D}\right)\right] \tag{8.2}$$

where $R_{in} = R(0) > D/c$. It is not difficult to check that $R(t)$ given by (8.2) is a concave curve; therefore, it describes species spread with increasing speed.

However, a closer inspection of Eq. (8.2) shows that it gives a pattern of spread different from that observed for Japanese beetle. While in the latter case the species range is increasing gradually, what we have in case of Eq. (8.2) looks more like a change between two constant-rate asymptotics with different speed, one for small t,

$$R \simeq R_{in} + \left(c - \frac{D}{R_{in}}\right)t \tag{8.3}$$

and the other for large t,

$$R \simeq const + ct \ . \tag{8.4}$$

The area radius vs time as given by Eq. (8.2) along with asymptotics (8.3) and (8.4) is shown in Fig. 8.3. (Note that the value of the constant in (8.4) is unknown but it is the slope of the line that is important, not its exact position.) Thus, although a qualitatively similar scenario of species invasion has been observed for some avian invasion, e.g., for the invasion of the European starling and the house finch in North America (Okubo, 1988; Shigesada and Kawasaki, 1997), it is unlikely to be applicable to the Japanese beetle invasion. Also, it must be mentioned that Eq. (8.1) was actually derived under the assumption that the "width" of the population front, i.e., the transition region between the densely populated areas (behind the front) and the areas where the invasive species is still absent (in front of the front), is sufficiently narrow. Apparently, this is not always the case; an alternative scenario is given by a self-similar expansion when growth of the area is accompanied by growth of the front width.

An exact self-similar solution describing the early stage of invasion has been obtain in Section 4.2 based on the modified Fisher equation

$$\frac{\partial u}{\partial t} = \left(\frac{\partial^2 u}{\partial r^2} + \frac{\eta}{r}\frac{\partial u}{\partial r}\right) + f(u) \tag{8.5}$$

(in dimensionless variables) where the generalized logistic population growth $f(u) = f_\gamma(u)$ is changed to logarithmic growth $f(u) = \bar{f}(u)$; see (4.35) and

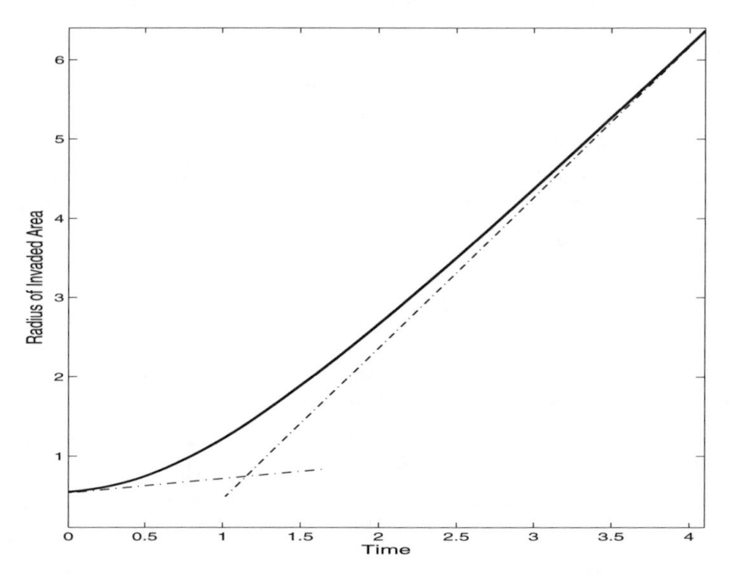

FIGURE 8.3: Invasion at increasing rate due to the interplay between the space dimension and the impact of the initial conditions. The solid curve gives the radius $R(t)$ of invaded area as given by (8.2); the dashed-and-dotted lines show the small-time and the large-time asymptotics.

(4.37). According to the solution, the radius of the invaded area is given by the following equation:

$$R(t) = \left[\left(4 + R_{in}^2\right) e^t - 4\,\right]^{1/2} , \tag{8.6}$$

cf. (4.46) and also see (4.47) and (4.40). It was shown that, although for large time it exhibits an unrealistic exponential growth, for small t the predicted value of area radius complies very well with that obtained in the biologically realistic case $f = f_\gamma$.

Now, the point of interest is whether that exact solution is applicable to Japanese beetle invasions. Note that (8.6) gives the radius versus time in dimensionless variables; to calculate it in original dimensional variables, we need to know the value of the scales for R and t. For that purpose, we first need to estimate the values of parameters, i.e., diffusivity D and the radius l of the originally invaded domain. Let us mention that, usually, obtaining the value of diffusivity for a given species is a difficult problem, cf. Kareiva (1983). In the case of the Japanese beetle spread, however, we can make use of the fact that, during its later stage (i.e., 1925–41), the invasion took place with a constant speed. Taking into account that for these years the radius of the invaded area is already rather large (on the order of one hundred km), the formula for the plane traveling wave speed seems to be applicable, i.e., $c = 2\sqrt{D/\tau}$ where τ is the characteristic time of population multiplication. It is readily seen from the data shown in Fig. 8.2 that the speed of the Japanese beetle invasion during these years can be estimated as 5.5 km/year. Then, assuming that $\tau \approx 0.033$ year, we arrive at $D \approx 0.25$ km^2/year. We want to mention that we are not aware of any existing estimates of τ and D for Japanese beetle; thus it seems impossible to check these values directly. It should be noted, however, that the values of τ and D accepted above are of the same order as the characteristic time and diffusivity for other insect species.

In order to obtain the value of l, one needs to know details of the initial distribution of an invasive species. This also brings a problem because in most cases regular observations on species spread start not immediately from the moment when the species is discovered in a new environment but only after a certain time. This concerns the spreading of the Japanese beetle as well; the data about its initial spatial distribution are somewhat contradictory (note that here by the "initial distribution" we mean the distribution at the time when the species is observed for the first time). Choosing $l = 0.5$ km (which seems to be consistent with the rest of the data), the scaling factors for R and t become 0.5 km and $l^2/D = 1$ year correspondingly.

The last task is the choice of the value for parameter R_{in}. According to its meaning, see the comments below Eq. (4.39), this parameter takes into account not only details of the initial species distribution but also details of the method of the field observations, particularly, the value of the threshold density (i.e., the density below which the population cannot be detected). The lack of information makes it impossible to obtain a reliable estimate for R_{in}. Fortunately, the behavior of $R(t)$ given by (8.6) appears to be rather robust to variations of this parameter. Fig. 8.2 shows the radius of the area invaded by Japanese beetle versus time calculated according to (8.6) for the values of τ, D and l as chosen above and for three different values of R_{in} (solid curves). Thus, exact solution (8.6) provides an apparently good description of the early stage of the invasion.

Interestingly, prediction made by using Eq. (8.5) with a more realistic growth function $f(u) = f_\gamma(u)$ (see (4.35)) appears to be in a worse agreement with the data because it gives a much lower rate of the area growth. The dashed-and-dotted line in Fig. 8.2 shows the numerical solution for the problem (4.33), (4.35), (4.39) obtained for $\gamma = 0.033$ (in accordance with the estimates for D, τ and l made above) and $R_{in} = 1$ (the solutions for the other two values of R_{in} used in Fig. 8.2 are not shown because all the three curves nearly coincide with each other).

In conclusion, we want to mention that the good agreement between the theory and the observations is, to a certain extent, subject to the appropriate choice of parameter values. Although small variations of the parameters (within a few percent of the values used for Fig. 8.2) do not break the agreement, for essentially different values the discrepancy between the solution (8.6) and the field data may become significant. Also, the field data itself give rise to some questions, e.g., about the discontinuity between the parts corresponding to the constant-rate and the increasing-rate stages of the Japanese beetle spreading. This discontinuity may have been caused by a variety of reasons, e.g., by a modification of the monitoring criteria. It can hardly be expected that these complicated and rather obscure circumstances could be taken into account in terms of our conceptual model. Still we think that a very good coincidence between the theory and the field data shown in Fig. 8.2 can be regarded as an indication that the exact solution obtained in Section 4.2 is not just a mathematical toy but can have a variety of ecological applications.

8.2 Mount St. Helens recolonization and the impact of predation

There is now considerable evidence coming both from empirical studies and from theoretical research that predation can affect the invasion rate and essentially modify the whole scenario of species spread (Dunbar, 1983, 1984; Sherratt et al., 1995, 1997; Fagan and Bishop, 2000; Owen and Lewis, 2001; Fagan et al., 2002; Petrovskii et al., 2002; Torchin et al., 2003). One of the most refined examples of predation impact was recently obtained in the field studies on vegetation recolonization patterns at Mount St. Helens, a volcano situated in the state of Washington, USA. Its eruption in 1980 created a vast area free from any plant or animal species and a large outer zone where only very few of them survived in some scarce natural shelters. The primary succession started soon after the eruption (del Moral and Wood, 1993) and has been going up to date, and the area became a huge research site for studying early ecosystem development.

Among the first species coming back to the area were lupines, *Lupinus lepidus*. This species has been attracting significant attention ever since in order to better understand its ecophysiology and its contribution to soil formation and successional dynamics, e.g., see Halvorson et al. (1992) and Braatne and Bliss (1999). In particular, Fagan and Bishop (2000) studied the spatial aspects of lupine recolonization. They found that the species range had been increasing linearly with time during the first several years before it turned to a slowly decelerating regime which is described by a curve close to a straight line but with a lower slope (see Fig. 8.4). Remarkably, the time of this change in the range growth rate coincided with re-appearance of some herbivorous species. Apparently, that made a solid ground for linking the slower growth of lupine range to the impact of its consumers.

The proven relation between invasion rates and predation has yet left many issues unclear. It is not fully clear how significant the effect of predation can be, i.e., whether it can block species invasion or turn invasion into species retreat. Although Fagan and Bishop (2000) estimated that there should be the

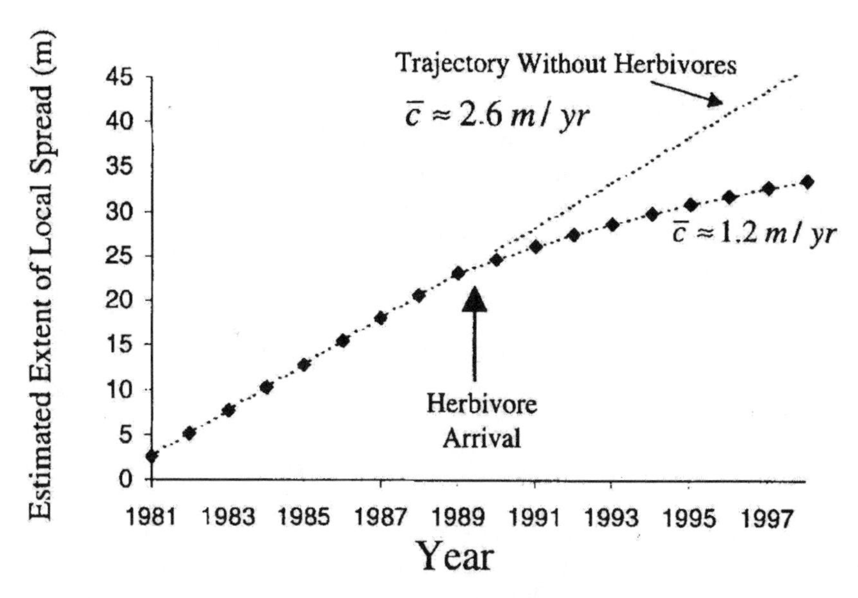

FIGURE 8.4: Advance of the boundary of the lupine range in the course of its recolonization over Mount St. Helens area (with permission from Fagan and Bishop, 2000).

turning point in the lupine range growth, the accuracy of their prediction does not seem to be high enough. It remains largely obscure what the biological requisites for the impact of predation to occur are, i.e., whether it is a common phenomenon or it takes place for a certain type of prey or predator. Finally, it is not clear what is the minimum modeling framework which is necessary to account for this phenomenon.

From the modeling perspective, it seems that the most natural approach to describe predator-prey interaction should include at least two dynamical variables. In case of a diffusion-reaction model, they would be prey density u and predator density v and the model should include two equations, respectively. However, it appears that a useful insight into the impact of predation on the invasion rate can be done already in terms of a single-species model.

Let us consider the following equation:

$$u_t(x,t) = Du_{xx} + F(u) - \sigma v u \tag{8.7}$$

where the last term in the right-hand side accounts for predation. Apparently, the model is incomplete and we should either add one more equation or make certain simplifying assumptions about the predator density. Here we assume that $v = v_0 = const$. Note that, although this is, of course, a rather restrictive assumption, it is not totally unrealistic. It may correspond to a homogeneously distributed immobile slowly growing predator; for instance, a few species with similar properties can be identified in marine benthic ecosystems.

The impact of predation then depends on the properties of the growth rate function $F(u)$. We distinguish between two qualitatively different cases, i.e., the case of logistic growth and the case of the strong Allee effect parameterized by the square and cubic polynomials, respectively. It is readily seen that, in both of these cases, predation slows down the spread of prey, although with essentially different consequences for the population dynamics.

Namely, in the case of logistic growth, from (8.7) we obtain:

$$u_t(x,t) = Du_{xx} + \alpha u \left(1 - uK^{-1}\right) - \sigma v_0 u \; . \tag{8.8}$$

Equation (8.8) can be written in a form identical to that of the absence of predation:

$$u_t(x,t) = Du_{xx} + \tilde{\alpha} u \left(1 - u\tilde{K}^{-1}\right) \tag{8.9}$$

where $\tilde{\alpha} = \alpha - \sigma v_0$ and $\tilde{K} = K(1 - \sigma v_0/\alpha)$. Therefore, the minimum value of the traveling wave speed is $c = 2\sqrt{D(\alpha - \sigma v_0)}$ and, obviously, it decreases as the intensity of predation increases, i.e., with an increase in v_0 and/or σ.

Note that an increase in the predation intensity has also a "global" impact on the population dynamics so that the carrying capacity decreases significantly with an increase in σv_0 (see Fig. 8.5). In particular, invasion blocking is reached for $\sigma v_0 = \alpha$ and for these values $\tilde{K} = 0$. Thus, in the case of logistic growth invasion blocking is possible only by means of total wipeout of the invasive species.

However, blocking and retreat of invasive species without its extirpation appears to be possible if we assume that its growth is damped by the strong Allee effect. In this case, Eq. (8.7) turns to

$$u_t(x,t) = Du_{xx} + \omega u(u - u_A)(K - u) - \sigma v_0 u \; . \tag{8.10}$$

In the absence of predator, the traveling population front propagates with the speed

$$c = \sqrt{\frac{D\omega}{2}} \cdot \left(1 - \frac{2u_A}{K}\right) ; \tag{8.11}$$

see Section 2.1.

Evidently, Eq. (8.10) can be written in the same form as in the absence of predator, i.e., as

$$u_t(x,t) = Du_{xx} + \omega u(u - \hat{u}_A)(\hat{K} - u), \tag{8.12}$$

where

$$\hat{u}_A = \frac{1}{2}\left[(K + u_A) - \sqrt{(K + u_A)^2 - 4\left(u_A K + \frac{\sigma v_0}{\omega}\right)}\,\right], \tag{8.13}$$

$$\hat{K} = \frac{1}{2}\left[(K + u_A) + \sqrt{(K + u_A)^2 - 4\left(u_A K + \frac{\sigma v_0}{\omega}\right)}\,\right]. \tag{8.14}$$

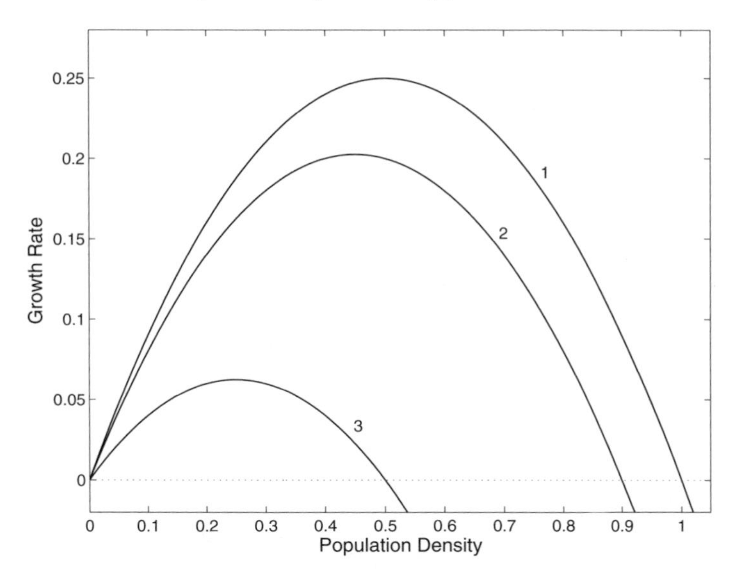

FIGURE 8.5: A sketch of prey population response to an increase in predation in the absence of the Allee effect. Curves 1, 2 and 3 show the growth rate versus population density for $\sigma v_0 = 0,\ 0.1$ and 0.5, respectively, cf. (8.8). Other parameters are $K = \alpha = 1$.

Since the growth function in Eq. (8.12) is still given by a cubic polynomial, the equation (8.11) for the front speed is applicable subject to the change $u_A \to \hat{u}_A,\ K \to \hat{K}$. Any increase in predation intensity σv_0 leads to an increase in $\hat{u}_A$ and to a decrease in $\hat{K}$ (see Fig. 8.6); therefore, it results in a decrease in the invasion speed.

From $c = 0$ we obtain the condition of wave blocking as $2\hat{u}_A = \hat{K}$ which, after some algebra, takes the following form:

$$\sigma v_0 = \left[\frac{2}{9}(1+\beta)^2 - \beta\right]\omega K^2 \tag{8.15}$$

where $\beta = u_A/K$. Predation will lead to retreat of the invasive species when its intensity σv_0 is higher than that given by (8.15). Note that blocking and retreat of the invasive species now takes place without its global wipeout, cf. Fig. 8.6. Although an increase in the predation intensity does result in a decrease in the prey carrying capacity, it remains well above zero until it suddenly disappears for $\hat{u}_A = \hat{K}$.

The critical relation given by Eq. (8.15) is shown in Fig. 8.7 by the thick curve. We want to emphasize that, in the absence of predation, species retreat occurs only for $\beta > 0.5$ and the whole domain on the left of the vertical line $\beta = 0.5$ would correspond to species invasion. Thus, species retreat for parameter values from the domain above the thick curve and on the left of the vertical dotted line must be essentially attributed to the impact of predation.

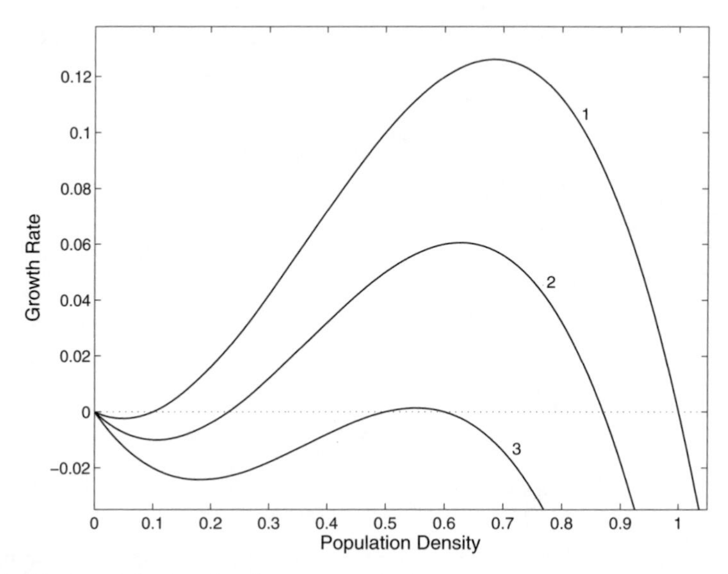

FIGURE 8.6: A sketch of prey population response to an increase in predation under the impact of strong Allee effect. Curves 1, 2 and 3 show the growth rate versus population density for $\sigma v_0 = 0,\ 0.1$ and 0.2, respectively, cf. (8.10). Other parameters are $K = \omega = 1,\ u_A = 0.1$.

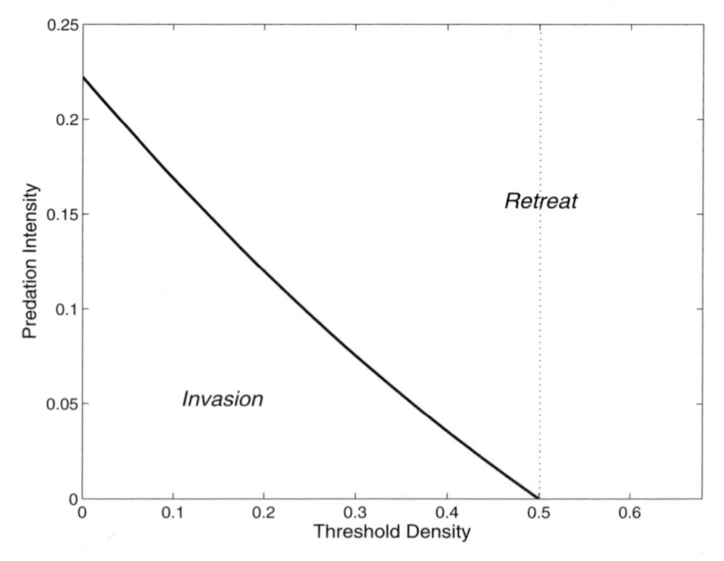

FIGURE 8.7: A map in the parameter plane $(\beta, \sigma v_0)$ of the single-species model (8.10) for $K = \omega = 1$.

The above conclusions have been made based on the assumption that predator density remains homogeneous in space and does not change with time. Besides the restrictions that this assumption imposes on the predator motility and growth (see the comments below Eq. (8.7)), it also implies a more specific presumption regarding the type of the predator. Namely, let us notice that $v(x,t) = v_0 = const$ actually means that the predator dwells all over the domain, i.e., in the areas that are already occupied by the invading prey as well as in the areas that are not yet invaded. The fact that the predator can survive in the absence of the prey may be interpreted that it is a *generalist* predator, i.e., it can feed on many different species. An alternative case is given by a *specialist* predator when it can only feed on one particular species. Apparently, the assumption $v(x,t) = v_0 = const$ is by no means applicable to a specialist predator. In order to give this case a full consideration, we have to analyze a predator-prey system consisting of two partial differential equations. This will be done below; however, it seems that a certain insight can still be made in terms of a single-species model.

For that purpose, we consider the case when the relation between the given predator and its prey implies, in the large-time limit, their steady coexistence at a certain equilibrium density, (u_0, v_0). We then assume that the ratio of predator and prey densities remains constant during their approach to the equilibrium, $v(x,t)/u(x,t) = v_0/u_0 = \varrho$. Correspondingly, Eq. (8.7) turns to

$$u_t(x,t) = Du_{xx} + F(u) - \sigma\varrho u^2 \ . \tag{8.16}$$

In the spatial perspective, the assumption $v = \varrho u$ means that both $u(x,t)$ and $v(x,t)$ show the same rate of decay at the tail in front of the population front. It should also be mentioned that, besides the case of a specialist predator, the linear relation between the prey and predator densities may reflect the fact that, in front of the front of invasive prey, there is no other species at all. The latter situation directly applies to Mount St. Helens recolonization.

It is readily seen that, since the last term in (8.16) does not alter the per capita population growth rate at small u, in the case of logistic growth predation by a specialist predator does not modify the rate of the prey invasion. The situation is different for a population with the Allee effect. In this case, instead of Eq. (8.10) we have

$$u_t(x,t) = Du_{xx} + \omega u(u - u_A)(K - u) - \sigma\varrho u^2 \tag{8.17}$$

which can be written as

$$u_t(x,t) = Du_{xx} + \omega u(u - \tilde{u}_A)(\tilde{K} - u) \tag{8.18}$$

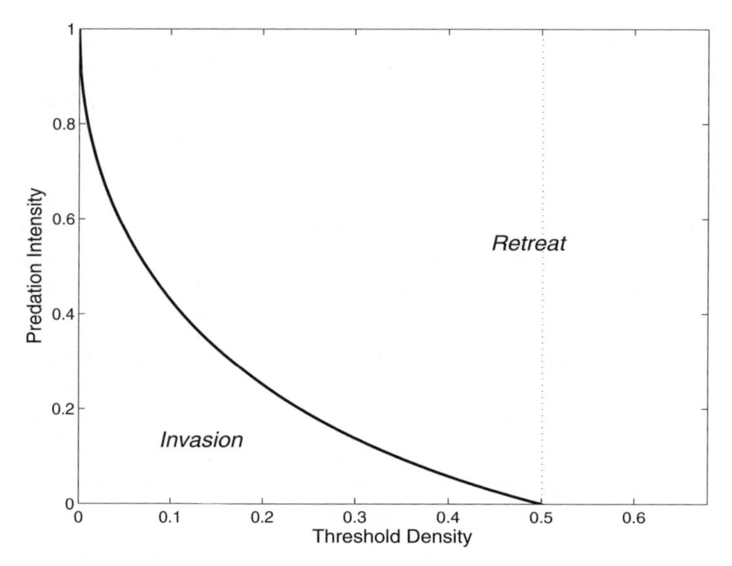

FIGURE 8.8: A map in the parameter plane $(\beta, \varrho\sigma)$ of the single-species model (8.17) for $K = \omega = 1$.

where

$$\tilde{u}_A = \frac{1}{2}\left[\left(K + u_A - \frac{\sigma\varrho}{\omega}\right) - \sqrt{\left(K + u_A - \frac{\sigma\varrho}{\omega}\right)^2 - 4u_A K}\,\right], \quad (8.19)$$

$$\tilde{K} = \frac{1}{2}\left[\left(K + u_A - \frac{\sigma\varrho}{\omega}\right) + \sqrt{\left(K + u_A - \frac{\sigma\varrho}{\omega}\right)^2 - 4u_A K}\,\right]. \quad (8.20)$$

From the condition of the wave blocking, i.e., $2\tilde{u}_A = \tilde{K}$, we arrive at the following critical relation between the parameters:

$$\sigma\varrho = \left(1 + \beta - 3\sqrt{\frac{\beta}{2}}\right)\omega K \quad (8.21)$$

where $\beta = u_A/K$.

The relation (8.21) is shown in Fig. 8.8 by the thick curve. As well as in the case of a generalist predator, the impact of specialist predator significantly decreases the parameter domain allowing for species invasion. Since in the absence of predator the whole area on the left of the vertical (dotted) line $\beta = 0.5$ would correspond to invasion, species retreat that takes place for parameters from above the thick curve is essentially the result of predation.

Now, we proceed to consideration of the predator impact in terms of the full predator-prey system:

$$u_t(x,t) = D_1 u_{xx} + F(u) - r(u)uv \ , \tag{8.22}$$
$$v_t(x,t) = D_2 v_{xx} + \kappa r(u)uv - g(v)v \ . \tag{8.23}$$

System (8.22–8.23) was analyzed in application to species invasion by Owen and Lewis (2001). They showed that, in the case that prey exhibits a generalized logistic growth, cf. (1.8–1.10), predation cannot slow down prey invasion. This is consistent with the results of our analysis for $v = \varrho u$, i.e., in the case of a specialist predator. They also showed that predation does slow down the prey invasion if prey growth is damped by the Allee effect; however, for this case their study was restricted to $D_1 \ll D_2$.

In order to address the situation when D_1 and D_2 are of the same order (ecological relevance of this case is discussed in Section 6.1.2), we put $D_1 = D_2 = D$ and try to make use of the exact solution that has been obtained in Section 6.1 for a particular case of (8.22–8.23):

$$u_t = u_{xx} - \beta u + (\beta + 1)u^2 - u^3 - uv \ , \tag{8.24}$$
$$v_t = v_{xx} + kuv - mv - \delta v^3 \ . \tag{8.25}$$

It was shown that, under additional constraints imposed on parameter values,

$$m \ = \ \beta \ , \qquad k + \frac{1}{\sqrt{\delta}} \ = \ \beta + 1 \ , \tag{8.26}$$

Eqs. (8.24–8.25) have exact solution (6.28–6.29) which describes, in the large-time limit, a traveling front propagating with the speed

$$n_2 = \frac{1}{2\sqrt{2}} \left(k - 3\sqrt{k^2 - 4m} \right) . \tag{8.27}$$

Remarkably, the direction of propagation of the population front can be different, cf. (6.33). While in the case $n_2 < 0$ the population front propagates to the region with low population density which corresponds to species invasion (for appropriately chosen conditions at infinity), in the case $n_2 > 0$ the front propagates to the region with high population density and thus it corresponds to species retreat. Moreover, the case $n_2 > 0$ (see also (6.33)) actually corresponds to an interplay between two different mechanisms; each of them can reverse the traveling front. One of these mechanisms is associated with the Allee effect and the other is related to the impact of predation.

In order to distinguish between the two mechanisms, let us first mention that under constraints (8.26) variables u and v become proportional to each other, $v = u/\sqrt{\delta}$, and system (6.8–6.9) is virtually reduced to one equation:

$$u_t = u_{xx} - \beta u + \left(\beta + 1 - \frac{1}{\sqrt{\delta}} \right) u^2 - u^3 \ . \tag{8.28}$$

We want to emphasize that the linear relation between the species densities here is not an additional assumption; it arises as a consequence of the constraints (8.26) which have a clear biological meaning, cf. Section 6.1.2.

In the single-species model with the population growth described by a cubic polynomial of a general form, the speed of the front propagation is given by the following equation (Murray, 1993; Volpert et al., 1994):

$$n = -\frac{1}{\sqrt{2}}\left(s_2 + s_0 - 2s_1\right) \tag{8.29}$$

(in dimensionless variables) where $s_0 \le s_1 \le s_2$ are the roots of the polynomial and minus corresponds to the choice of the conditions at infinity as $u(-\infty, t) = s_0$, $u(+\infty, t) = s_2$. In the prey-only limit of the system (8.24–8.25), i.e., in the case $v \equiv 0$, we obtain:

$$n = -\frac{1}{\sqrt{2}}\left(1 - 2\beta\right), \tag{8.30}$$

so that the front propagates toward the region where $u \approx 0$ for $\beta < 0.5$ (species invasion) and toward the region where $u \approx 1$ for $\beta > 0.5$ (species retreat); $\beta = 0.5$ corresponds to the front with zero speed. Note that β is a dimensionless measure of the Allee effect; thus, an increased Allee effect can turn invasion to retreat.

Now, what can change in the presence of predator, i.e., in case $v \neq 0$ identically? Having applied the comparison theorem for nonlinear parabolic equations (cf. Section 7.4) to Eq. (6.8), it is readily seen that predation cannot turn retreat back to invasion. However, the impact of predation can turn invasion to retreat even when the invasion would be successful in the absence of predator. In the predator-prey system (8.24–8.25), the "turning" relation between the parameters is:

$$k = 3\sqrt{\frac{m}{2}}, \tag{8.31}$$

cf. (6.33), so that the values of k greater than the one given by Eq. (8.31) correspond to species invasion. Smaller k corresponds to species retreat.

Allowing for relations (8.26), the condition of species retreat takes the following form:

$$\frac{1}{\sqrt{\delta}} > \beta + 1 - 3\sqrt{\frac{\beta}{2}}. \tag{8.32}$$

In the case $\beta > 0.5$, inequality (8.32) describes species retreat resulting from a joint effect of two factors, i.e., increased Allee effect (large β) and the impact of predation. Note that, as it can be seen from comparison between (8.27) and (8.30), the speed of retreat appears to be higher in the presence of predator. In the case $\beta < 0.5$, however, inequality (8.32) describes species retreat which must be essentially attributed to the impact of predation because in the absence of predator the condition $\beta < 0.5$ always leads to species invasion.

Note that the right-hand side of relation (8.32) coincides, up to parameters re-scaling, with the right-hand side of Eq. (8.21). This is not surprising because (8.21) is obtained under the assumption that the species densities are linearly related. Parameter δ thus serves as a specific measure of predation intensity, and relation (8.32) has a clear ecological interpretation: a weak predator (= high predator mortality, large δ) makes prey invasion possible, a strong predator (= low predator mortality, small δ) blocks invasion and turns it into retreat, cf. the domains below and above the thick curve in Fig. 8.8.

Thus, we have shown by using different models that predation can have a diverse effect on species spread depending on the type of the density-dependence in the invasive species growth (i.e., with or without the Allee effect), the type of predator (generalist or specialist) and/or on some details of species spatial distribution, i.e., whether prey invades into the area where the predator is already established (which implies that it has an alternative source of food) or the predator is only present behind the propagating front. In general, predation tends to decrease the rate of prey invasion but blocking and retreat can only be possible if prey growth is damped by the Allee effect. Although we cannot directly compare these theoretical results with the field data on Mount St. Helens recolonization because of lack of important information about parameter values and the type of prey (lupines) growth, qualitatively, they seem to be in a good agreement.

8.3 Stratified diffusion and rapid plant invasion

As it has been already mentioned a few times throughout this book, the rates of geographical spread of invasive species sometimes appear to be significantly higher than that predicted by the "classical" diffusion-reaction models. In particular, higher rates of invasion and colonization have often been reported for plant species based both on ongoing ecological observations and on historical data; see Clark et al. (1998) for examples and further reference. There have been several attempts to explain this phenomenon of "rapid invasion" and to develop a theory and/or a modeling framework that would take it into account. In particular, Kot et al. (1996) ascribed rapid plant migration to the impact of non-Gaussian diffusion. They showed that in case the seed dispersal (mostly driven by wind) from a single parent plant is described by a fat-tailed distribution function, then the invasion speed can exceed the standard value $c = 2\sqrt{D\alpha}$ significantly.

Although the theory by Kot et al. (1996) indeed provides a plausible explanation for rapid plant invasion, there remain some questions that have not been properly addressed yet. The theory by Kot et al. (1996) is essentially based on the assumption that the rate of the dispersal kernel decay in the large-distance limit is slower than $\exp(-r^2)$; however, whether the seed dispersal is actually fat-tailed or not has never been proved by field observations. (Note that it is the asymptotical behavior that matters and thus any particular set of data obtained at finite, albeit large, distances can hardly be used as a proof.) Also, it seems that a purely mechanistic approach relating the properties of the seed dispersal solely with peculiarities of turbulent wind mixing somehow underestimates the importance of biological factors. If we readily accept that the temporal dynamics of communities is to a large extent affected by inter-species interactions, then why do we neglect the spatial aspect of these interactions?

Remarkably, there exists another example of invasion at a higher speed, a phenomenon called "stratified diffusion" (Hengeveld, 1989; Shigesada et al., 1995). Geographical spread of some avian species takes place by following a two-phase scenario. At the beginning of the spread, the growth of the species range occurs in accordance with usual local diffusion and is well described by the standard diffusion-reaction models. At a later stage, however, the rate of the range growth increases considerably and can be as much as several times faster than that determined by the local diffusion. The spread at a higher rate has been attributed to the "jump dispersal" when a group of birds migrates and establishes a new colony far away from the main range. The jump dispersal mode has been noticed for several bird and insect species (Mundinger and Hope, 1982; Okubo, 1988; Liebhold et al., 1992; Sharov and Liebhold, 1998). The fact that invasion at a higher speed takes place only after some time after the beginning of geographical spread is regarded as a consequence of density-dependence so that the jump mode is "turned on" only when the population density within the species range grows high enough.

In this section, we make an attempt to explain rapid plant invasion by means of linking it to the stratified diffusion. We will show that there may be an intrinsic relation between these two phenomena and that the jump dispersal of avian species may result in a significant increase in the rates of some plant invasions. Mathematically, in order to describe the stratified diffusion we use the exactly solvable model developed in Section 4.1 which relates it to small-scale density-dependent migrations within the expanding population.

It should be mentioned that applicability of the diffusion-reaction equations to population dynamics under the impact of stratified diffusion may be not so obvious and must be justified. At first sight, it may seem that they are at variance with ecological observations. The issue of concern is the structure of the front, i.e., the transition region between the densely populated areas (behind the front) and the areas where the invasive species is fully or virtually absent (in front of the front). Diffusion-reaction equations normally predict a gradual monotone change in population density in the direction across the front. Instead, the transition region consists of an ensemble of colonies, or

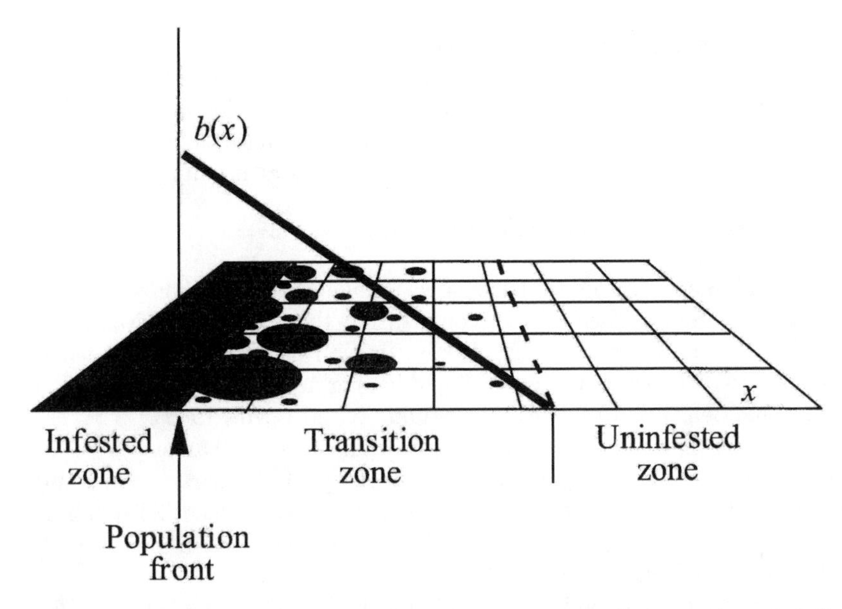

FIGURE 8.9: The structure of the transition zone between invaded (infested) and uninvaded (uninfested) regions in the case that species spread is affected by the stratified diffusion. Here $b(x)$ sketches population density versus space through the transition zone according to the modified definition; see details in the text (with permission from Sharov and Liebhold, 1998).

"patches," of various shape and size. A typical situation is sketched in Fig. 8.9. The population reaches a considerable density inside these patches and is absent in the areas between the patches. Note that the density of the patches themselves tends to decrease as the distance from the species' main range increases.

In order to resolve this seeming contradiction, let us recall the origin of the diffusion equation and the presuppositions that create the framework for its application, cf. Section 2.1. The whole idea of describing the dynamics of given species in terms of its *population density* implies an averaging over a certain area, or "window". It is this averaging procedure that allow us to describe the phenomenon of stochastic origin, i.e., diffusion, by means of deterministic equations. The size or radius of the window must be small enough in order to relate its position to a single point in space but it also should be large enough to contain a sufficiently large number of individuals of given species in order to exclude fluctuations of purely stochastic nature. Now, we can apply a similar approach to the patchy distributed population. Choosing the averaging window in the form of a narrow long stripe oriented along the front, we arrive at the population density that exhibits a gradual change in the direction across the front (cf. the thick line in Fig. 8.9), in the manner congenial to what is predicted by diffusion-reaction equations.

Our model is as follows. Let $u(x,t)$ be the density of a plant species $\mathcal{U}$. For convenience, we regard $\mathcal{U}$ as a tree species, although this is not a principal constraint. We consider the dynamics of species $\mathcal{U}$ in an infinite domain and assume that density u has the following conditions at infinity:

$$u(x \to -\infty, t) \;=\; 0, \qquad u(x \to \infty, t) \;=\; 1. \tag{8.33}$$

Conditions (8.33) implies that the species range has a boundary situated at a certain $\bar{x}$. Species invasion or colonization takes place if, with each new generation, $\bar{x}$ moves to the left. Redistribution of species $\mathcal{U}$ in space takes place due to dispersal of its seeds which, having neglected the impact of other species, happens due to wind mixing (Okubo and Levin, 1989).

In practice, however, wind mixing is not the only mechanism of seeds dispersal. Seeds are eaten by animals and birds near the parent plant and, in due time, are egested after being transported to a new place. Although some of the seeds are likely to be destroyed by the process of digestion, others may still retain their germinative ability.

Now, we consider an avian species $\mathcal{V}$ for which the seeds of species $\mathcal{U}$ is an essential source of food. We denote the density of species $\mathcal{V}$ as $v(x,t)$ and consider the conditions at infinity the same as (8.33):

$$v(x \to -\infty, t) \;=\; 0, \qquad v(x \to \infty, t) \;=\; 1. \tag{8.34}$$

Apparently, the relation between the species $\mathcal{U}$ and $\mathcal{V}$ is of resource-consumer or predator-prey type. A model providing the full mathematical description of this system's dynamics would consist of two equations, respectively. However, we can make use of the observation that higher density of resource (or prey) often leads to a higher density of consumer (or predator). Since species $\mathcal{V}$ feeds on species $\mathcal{U}$, we assume that within the range of species $\mathcal{U}$ their densities are linearly dependent:

$$v(x,t) = \varrho u(x,t). \tag{8.35}$$

Note that this assumption does not necessarily mean that species $\mathcal{V}$ is a specialist consumer (cf. "specialist predator," Section 8.2). Alternatively, it could imply that species $\mathcal{V}$ dwells in a mono-species forest created by $\mathcal{U}$ and the fact that $\mathcal{V}$ mostly feeds on $\mathcal{U}$ would simply reflect the lack of choice.

We then assume that species $\mathcal{V}$ exhibits two-mode dispersal, i.e., the short-distance mode due to local diffusion and the long-distance mode due to migration. The colonies created by the groups of migrating birds are outside of the main range of species $\mathcal{U}$ and thus relation (8.35) does not seem to be applicable. However, we recall that, before migrating, the birds have been feeding on the seeds of $\mathcal{U}$ and it is likely that some of them are transported to the place where a new colony is formed. Moreover, the more birds have migrated the more seeds are transported; thus, we assume that outside of the species $\mathcal{U}$ range the density of $\mathcal{U}$ and $\mathcal{V}$ are linearly dependent as well but with a different coefficient:

$$u(x,t) = \frac{1}{\varrho_1}\, v(x,t). \tag{8.36}$$

A more careful analysis of this population system would require the consideration of different ϱ and ϱ_1. Here, in order to make an early insight into the impact of the avian-based seed transport, we assume that $\varrho = \varrho_1$. The system dynamics is then virtually described by a single variable $v(x,t)$ and by a single equation:

$$\frac{\partial v(x,t)}{\partial t} + \frac{\partial (Av)}{\partial x} = D\frac{\partial^2 v}{\partial x^2} + F(v)\ , \tag{8.37}$$

cf. (2.7), where $A = A(v)$ must be a monotonously growing function in order to take into account the fact that the migration intensity tends to increase as the population density increases (Hengeveld, 1989). Although the properties of Eq. (8.37) in a more general case remain to be investigated, for the case of specific parameterization given by a linear function for $A(v)$ and a cubic polynomial for $F(v)$, Eq. (8.37) has an exact solution describing propagation of the population front (see Section 4.1). The speed of the front is shown to be

$$q_2 = \beta\nu - \frac{1}{\nu} \tag{8.38}$$

where $\nu = 0.5(a_1 + \sqrt{a_1^2 + 8})$ and a_1 is the migration intensity (in dimensionless variables). Correspondingly, the speed accretion, i.e., the additional speed that the front acquires due to species migration, is given as

$$\Delta q = q_2 - \frac{(2\beta - 1)}{\sqrt{2}}\ . \tag{8.39}$$

It has then been shown that Δq is a monotonously growing function of a_1 so that, for a_1 being on the order of unity or greater, the front speed q_2 can considerably exceed the speed observed in the no-migration case (see Fig. 4.1).

Now, what can be the value of a_1 for an avian species exhibiting the jump dispersal mode? According to its definition,

$$a_1 = \frac{2A_1}{\sqrt{D\omega}} \tag{8.40}$$

(see the lines below Eq. (4.5)). Let us notice that $c = \sqrt{DK^2\omega/2}$ is the speed of invasion due to the usual diffusion-reaction mechanism (i.e., the short-distance mode) in the case that the alien species is damped by the Allee effect; see Eq. (2.20). Equation (8.40) then takes the form

$$a_1 = \sqrt{2}\,\frac{A_1 K}{c}\ . \tag{8.41}$$

The value of c can be obtained directly from relevant ecological data, without making separate estimates for D and ω. Field observations show that, typically, c lies between 3 and 12 km/year (Shigesada and Kawasaki, 1997); thus, in order to proceed further we take a characteristic value $c = 8$ km/year.

Note that, due to the conditions imposed at infinity, cf. (8.34), we are interested in the migration that goes against axis x, i.e., $A_1 < 0$. To estimate $|A_1|K$, let us mention that, due to its biological meaning, it can be written as follows:

$$|A_1|K = \Upsilon c_* \tag{8.42}$$

where c_* is the characteristic speed of bird's travel and Υ is the proportion of given avian population that exhibits the tendency to small-scale migration. Since the dynamics of avian populations is to a large extent affected by seasonality, we assume that Υ applies to a fixed term of one year. Because the value of Υ is unknown, we consider a hypothetical value $\Upsilon = 0.001$. Then, assuming $c_* = 100$ km/day, we obtain $|A_1|K = 100 \cdot 365 \cdot 0.001 = 36.5$ km/year. Finally, taking into account the sign of A_1, from (8.41), we arrive at $a_1 \approx -6.5$.

The threshold density β is very difficult to estimate but it is likely to be small, $\beta \ll 1$. The corresponding dependence of Δq on a_1 is then given by curve 1 in Fig. 4.1. Therefore, we obtain that $\Delta q \approx -2.8$ while the speed in the "no-migration" case is -0.71. Thus, the rate of species invasion under the jump dispersal mode appears to be about five times higher compared to the invasion speed caused by the local diffusion. Note that this estimate is in a good agreement with field observations, cf. Shigesada and Kawasaki (1997), p.13 to 18.

Since in our model the densities of species $\mathcal{U}$ and $\mathcal{V}$ are linearly related (see (8.36)), fast expansion of the avian species range results in fast expansion of the plant species range. The rate of plant invasion due to the seed dispersal by wind normally varies between a few meters and a few dozen meters per year; in case the seed dispersal is enhanced by the avian-based transport, it can become many times higher. Moreover, our model actually predicts that the rate of plant invasion will increase with time when the jump dispersal mode of the corresponding avian species is turned on. Curiously, ecological data indeed show that invasion of some plant species may be going at an increasing rate.

In conclusion, we want to emphasize that the approach that we have used here is a very simple one and is based on a few strong assumptions. A more focused ecological study can and should be improved in many ways. The estimate of invasion rate that we obtained above is tentative and may only be applied to a specific ecological situation with caution. Apart from the model improvements, in order to obtain a more solid estimate one should possess more information about parameters β and Υ which is currently not available. Yet we think that the insight that we have made here based on our conceptual model clearly demonstrates that avian-based seed dispersal can change the pattern of plant invasion significantly and may be in some cases responsible for the rapid plant migration.

Chapter 9

Appendix: Basic background mathematics

9.1 Ordinary differential equations and their solutions

1. The equation

$$\tilde{G}\left(t, x(t), \frac{dx(t)}{dt}, \ldots, \frac{d^n x(t)}{dt^n}\right) = 0, \tag{9.1}$$

where $\tilde{G}$ is a certain function of $(n+1)$ arguments is called an *ordinary differential equation* (ODE) of the n-th order. Any function $x(t)$ that turns Eq. (9.1) to identity is called its solution. A solution that can be expressed via elementary or special functions is called a *solution in a closed form.* It is not always possible. Differential equations that have closed form solutions are often called *integrable.* In a more general case, a solution can be expressed via expansion into a series, e.g., power series, but not via elementary or special functions. Although a lot depends on the form of function $\tilde{G}$, a general observation is that the higher is the equation order the less likely it is integrable.

In case Eq. (9.1) can be written in the form

$$\frac{d^n x(t)}{dt^n} = G\left(t, x(t), \frac{dx(t)}{dt}, \ldots, \frac{d^{n-1} x(t)}{dt^{n-1}}\right), \tag{9.2}$$

it is called an ODE resolved with respect to the highest order derivative.

2. The following equation

$$\frac{d^n x(t)}{dt^n} + a_{n-1}\frac{d^{n-1} x(t)}{dt^{n-1}} + \ldots + a_1\frac{dx(t)}{dt} + a_0 x(t) + a = 0, \tag{9.3}$$

where a, a_0, $a_1, \ldots, a_{n-1}$ can depend on t but neither on x nor its derivatives, is called a *linear* ODE of the n-th order. The main property of a linear ODE is that the sum of any two of its solutions is also its solution.

In case $a = 0$ and all a_0, $a_1, \ldots, a_{n-1}$ are constant coefficients, the general solution of Eq. (9.3) can be written explicitly. It is readily seen that $\bar{x}(t) = \exp(\lambda t)$ is a solution of (9.3) provided that λ is a solution of the following algebraic equation:

$$\lambda^n + a_{n-1}\lambda^{n-1} + \ldots + a_1\lambda + a_0 = 0. \tag{9.4}$$

Solutions of Eq. (9.4) are called the *eigenvalues* of the ODE (9.3).

The form of the general solution depends on whether Eq. (9.4) has multiple solutions or all the eigenvalues are different. Let us consider these two cases separately.

(i) There are no multiple eigenvalues, i.e., if $i \neq k$ then $\lambda_i \neq \lambda_k$, $i, k = 1, \ldots, n$. In this case, the general solution has the form

$$x(t) = A_1 e^{\lambda_1 t} + A_2 e^{\lambda_2 t} + \ldots + A_n e^{\lambda_n t} , \tag{9.5}$$

where $A_1,\ A_2, \ldots, A_n$ are arbitrary constants.

(ii) Let the first m eigenvalues coincide and the rest of them be different, i.e., $\lambda_1 = \lambda_2 = \ldots = \lambda_m = \lambda$ and $\lambda_i \neq \lambda_k$ for any $i, k = m+1, \ldots, n$. Then, the general solution of Eq. (9.3) has the form

$$\begin{aligned} x(t) &= \left(A_1 + A_2 t + \ldots + A_m t^{m-1}\right) e^{\lambda t} \\ &\quad + A_{m+1} e^{\lambda_{m+1} t} + \ldots + A_n e^{\lambda_n t} . \end{aligned} \tag{9.6}$$

Cases (i) and (ii) can be readily extended to the general case when there are a few multiple eigenvalues.

The solution (9.5) or (9.6) is general in the sense that it contains all possible solutions of Eq. (9.3) and each particular solution can be obtained by choosing corresponding values of the constants $A_1,\ A_2, \ldots, A_n$. To obtain the constants A_i, Eq. (9.3) must be complemented with additional conditions and/or constraints. Most typically (but not necessarily), they are the conditions giving the value of variable x and its $(n-1)$ derivatives at a certain t_0:

$$\begin{aligned} x(t_0) &= x_0, \\ \left(\frac{dx}{dt}\right)_{t=t_0} &= x_0^{(1)}, \\ &\ldots, \\ \left(\frac{d^{n-1}x}{dt^{n-1}}\right)_{t=t_0} &= x_0^{(n-1)} . \end{aligned} \tag{9.7}$$

In case Eq. (9.3) is considered for $t > t_0$, relations (9.7) are called the *initial conditions.*

3. By means of introducing new variables,

$$x_1(t) = \frac{dx(t)}{dt}, \quad x_2(t) = \frac{d^2x(t)}{dt^2}, \quad \ldots, \quad x_{n-1}(t) = \frac{d^{n-1}x(t)}{dt^{n-1}}, \tag{9.8}$$

an ODE of the n-th order, cf. (9.2), is transformed to a system of n ordinary differential equations of the first order:

$$\begin{aligned} \frac{dx(t)}{dt} &= x_1(t) \ , \\ \frac{dx_1(t)}{dt} &= x_2(t) \ , \\ &\dots \\ \frac{dx_{n-1}(t)}{dt} &= G\left(t, x, x_1, \dots, x_{n-1}\right) . \end{aligned} \tag{9.9}$$

In some cases, the system (9.9) appears to be more convenient for analysis and understanding than the original equation (9.2) (see the next section). Note that the reverse transformation is also possible and a system of ODEs of the first order can be turned to a single high-order equation.

4. Since the models of mathematical ecology are usually based on *nonlinear* differential equations of either first or second order, their integrability is an issue of special interest. Below we briefly describe a few cases when the corresponding ODEs are likely to have solutions in a closed form and also give some hints regarding how those solutions can be found.

Remarkably, even in the simplest case of the first order equation,

$$\frac{dx(t)}{dt} = g(t, x) \ , \tag{9.10}$$

its solution cannot always be obtained in a closed form. Whether Eq. (9.10) is integrable or not depends on its right-hand side. There are a few cases when the solution can be found analytically. The most typical one is given by the situation when the variables can be separated, i.e., $g(t, x) = g_1(t)g_2(x)$. From (9.10), we then obtain:

$$\int \frac{dx}{g_2(x)} = \int g_1(t)dt \ . \tag{9.11}$$

In the case that the integrals are integrable, (9.11) leads to a solution of Eq. (9.10), at least, in an implicit form.

5. In the case of a nonlinear ODE of the second order,

$$\frac{d^2x(t)}{dt^2} = \tilde{g}\left(t, x, \frac{dx}{dt}\right) , \tag{9.12}$$

only relatively few cases can be solved in a closed form. Usually, these are the cases where the original equation can be reduced, e.g., by means of introduction of a new variable, to an ODE of the first order.

One particular case which often appears to be integrable corresponds to the situation when the equation does not contain x:

$$\frac{d^2x(t)}{dt^2} = g\left(t, \frac{dx}{dt}\right) . \tag{9.13}$$

Having introduced a new variable, $dx/dt = p(t)$, from (9.13) we obtain:

$$\frac{dp(t)}{dt} = g\left(t, p\right). \tag{9.14}$$

Integrability of Eq. (9.14) is then subject to the form of function g, cf. (9.10–9.11).

Another case when the order can be reduced arises if the right-hand side of Eq. (9.12) does not depend explicitly on the independent variable t:

$$\frac{d^2x(t)}{dt^2} = g\left(x, \frac{dx}{dt}\right). \tag{9.15}$$

Restricting the analysis to monotonous solutions, we can consider the derivative dx/dt as a function of x, i.e., $dx/dt = \psi(x)$ where ψ is to be determined. Then, instead of Eq. (9.15) we obtain

$$\psi\frac{\psi(x)}{dx} = g(x, \psi)\ . \tag{9.16}$$

Eqs. (9.14) and (9.16) are of the first order and thus are more likely to be integrable.

9.2 Phase plane and stability analysis

The following system of two ODEs of the first order,

$$\frac{dx(t)}{dt} = f(x, y)\ , \qquad \frac{dy(t)}{dt} = g(x, y)\ , \tag{9.17}$$

where f and g are certain functions, is called an *autonomous* in case f and g do not depend on variable t.

In an autonomous system, for each value of t the pair $x(t), y(t)$ can be conveniently interpreted as a point in the plane (x, y) which is called the *phase plane* of the system (9.17). Solutions of the system then correspond to curves or *trajectories* in the phase plane. Apparently, a solution of the system (9.17) with the initial conditions $x(0) = x_0,\ y(0) = y_0$ corresponds to a trajectory originating in the point (x_0, y_0). In a general case, when the right-hand side of equations depends also on t, the properties of the system trajectories can be very complicated. However, in the particular but important case of an autonomous system (see (9.17)), the trajectories' properties are somewhat simpler and can be studied in much detail, e.g., see Lefschetz (1963). Below we briefly recall those that can be helpful for understanding the content of this book.

The point $(\bar{x}, \bar{y})$ in the phase plane of the system (9.17) is called a *steady state* (also *equilibrium state, equilibrium point*) if

$$f(\bar{x}, \bar{y}) = 0, \quad g(\bar{x}, \bar{y}) = 0. \tag{9.18}$$

The set $\Omega_\epsilon = \{(x, y) : ([x - \bar{x}]^2 + [y - \bar{y}]^2)^{1/2} < \epsilon\}$ where $\epsilon > 0$ is called a ϵ-vicinity of the point $(\bar{x}, \bar{y})$.

A steady state $(\bar{x}, \bar{y})$ is called *asymptotically stable* if there exists $\epsilon > 0$ so that any trajectory originating in the ϵ-vicinity of $(\bar{x}, \bar{y})$ approaches the steady state in the large-time limit.

A steady state $(\bar{x}, \bar{y})$ is called *unstable* if, for any $\epsilon > 0$, there exists $(x_0, y_0) \in \Omega_\epsilon$ so that the trajectory originating in (x_0, y_0) will leave Ω_ϵ for sufficiently large t.

Regarding their stability, the steady states of the autonomous system (9.17) are classified into a few types. In order to arrive at this classification, we first consider the *linearized* system corresponding to (9.17):

$$\frac{dX(t)}{dt} = a_{11}X + a_{12}Y \ , \quad \frac{dY(t)}{dt} = a_{21}X + a_{22}Y \tag{9.19}$$

where

$$a_{11} = \left(\frac{\partial f}{\partial x}\right)_{(x_0, y_0)} , \quad a_{12} = \left(\frac{\partial f}{\partial y}\right)_{(x_0, y_0)} ,$$

$$a_{21} = \left(\frac{\partial g}{\partial x}\right)_{(x_0, y_0)} , \quad a_{22} = \left(\frac{\partial g}{\partial y}\right)_{(x_0, y_0)}$$

and the deviations from the steady state, $X = x - \bar{x}$ and $Y = y - \bar{y}$, are assumed to be small.

The following equation gives the eigenvalues of the linearized system:

$$\det(A - \lambda E) = 0 \tag{9.20}$$

where $A = (a_{ij})$ and E is the unit matrix.

The type of the steady state is defined according to the eigenvalue properties. Since (9.19) is a linear system, its solution is a linear combination of $e^{\lambda_1 t}$ and $e^{\lambda_2 t}$, cf. Section 9.1; thus, the steady state stability is subject to $\text{Re}\lambda_{1,2}$. The steady state $(\bar{x}, \bar{y})$ is stable if $\text{Re}\lambda_1 < 0$ and $\text{Re}\lambda_2 < 0$ and it is unstable if at least one of them is positive.

The details of trajectories' behavior in vicinity of a steady state also can be different depending on whether the eigenvalues are real or complex (see Fig. 9.1). In case both of them are real and have different signs, e.g., $\lambda_1 < 0 < \lambda_2$, the steady sate is called a *saddle*; the corresponding field of trajectories is shown in Fig. 9.1a. In case both eigenvalues are real and have the same sign, the steady state is called a *node* (see Fig. 9.1b). In case both eigenvalues are complex, the steady state is called a *focus* (see Fig. 9.1c). Apparently,

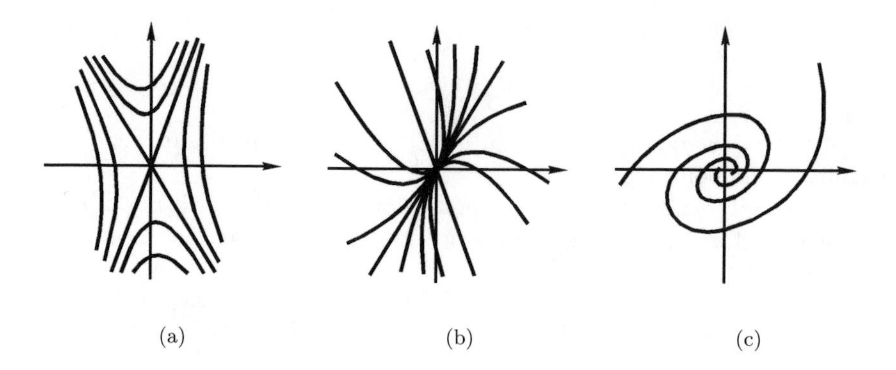

FIGURE 9.1: Examples of the phase plane structure in vicinity of (a) saddle, (b) node and (c) focus

saddle is always unstable while node and focus can be either stable or unstable. A complete classification also includes the cases when one or both of the eigenvalues are equal to zero; however, since these cases are structurally unstable we do not address them here.

9.3 Diffusion equation

The following partial differential equation

$$u_t = Du_{xx} \tag{9.21}$$

is usually called either the diffusion equation or the heat transfer equation because it was originally introduced to described these phenomena. Since diffusion is more congenial to species dispersal, in this book we always refer to Eq. (9.21) as a diffusion equation. Hence, $u(x,t)$ is the concentration of diffusing substance and D is the diffusion coefficient. We restrict our considerations to the case when Eq. (9.21) is defined in an unbounded space; thus, we are interested in the solution of the initial-value problem.

Let the initial condition to Eq. (9.21) be described by the Dirac δ-function:

$$u(x,0) = G\delta(x - x_0) \tag{9.22}$$

where

$$\delta(x - x_0) = 0 \ \forall\, x \neq x_0 \ \text{ and } \ \delta(x - x_0) = \infty \ \text{ for } x = x_0 \tag{9.23}$$

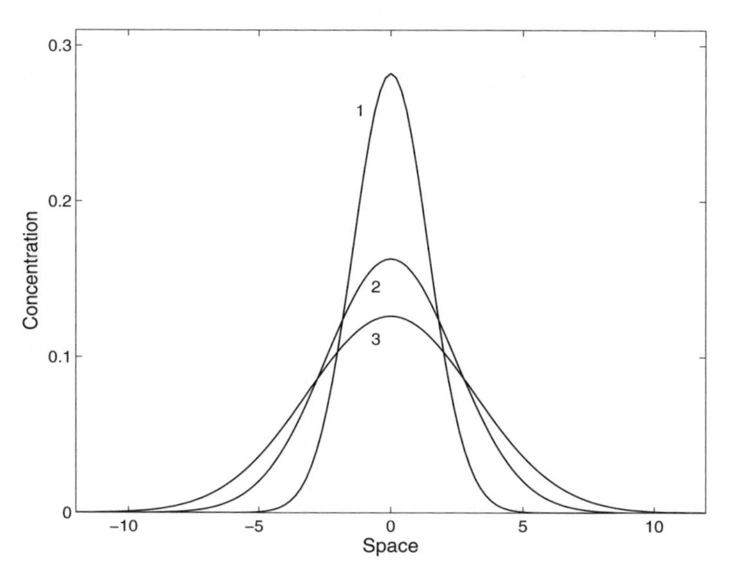

FIGURE 9.2: Fundamental solution (9.25) of the diffusion equation shown for $t = 1$ (curve 1), $t = 3$ (curve 2) and $t = 5$ (curve 3) for $D = G = 1,\ x_0 = 0$.

and, in spite of the singularity at $x = x_0$, function (9.23) is assumed to be integrable:

$$\int_{-\infty}^{\infty} \delta(x - x_0)dx = 1. \tag{9.24}$$

Clearly, from (9.22) we obtain that $\int_{-\infty}^{\infty} u(x, 0)dx = G$ so that G is the total amount of the diffusing substance released at the moment $t = 0$ at the position x_0.

Equation (9.21) with the initial condition (9.22) has the following solution:

$$u(x,t) = \frac{G}{\sqrt{4\pi Dt}} \exp\left(-\frac{(x - x_0)^2}{4Dt}\right). \tag{9.25}$$

Solution (9.25) is shown in Fig. 9.2.

The special solution (9.25) is called the *fundamental solution* because the solution of Eq. (9.21) with an arbitrary initial condition $u(x, 0) = \Phi(x)$ can be written as

$$u(x,t) = \frac{1}{\sqrt{4\pi Dt}} \int_{-\infty}^{\infty} \exp\left(-\frac{(x - x_0)^2}{4Dt}\right) \Phi(x_0)dx_0\ . \tag{9.26}$$

In order to better understand the properties of the diffusion equation and its solutions, let us now consider in somewhat more detail two examples where integration in (9.26) can be easily done.

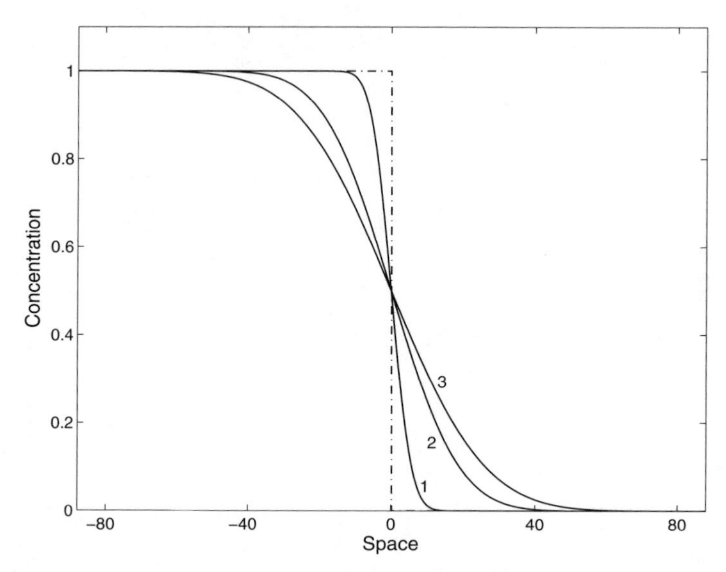

FIGURE 9.3: Solution (9.28) of the diffusion equation shown for $t = 10$ (curve 1), $t = 110$ (curve 2) and $t = 210$ (curve 3) for $D = U_0 = 1$; dashed-and-dotted line shows the initial conditions.

1. Semi-finite initial conditions. Consider

$$u(x,0) \;=\; U_0 \text{ for } x \le 0 \quad \text{and} \quad u(x,0) \;=\; 0 \text{ for } x > 0. \tag{9.27}$$

The corresponding solution of Eq. (9.21) is

$$u(x,t) = \frac{U_0}{2}\left[1 - \operatorname{erf}\left(\frac{x}{\sqrt{4Dt}}\right)\right], \tag{9.28}$$

where $\operatorname{erf}(z)$ is the error function,

$$\operatorname{erf}(z) = \frac{2}{\sqrt{\pi}}\int_0^z e^{-y^2}dy\ . \tag{9.29}$$

Solution (9.28) is shown in Fig. 9.3. It describes a self-similar distention of the boundary separating the domain where the diffusing substance is present in very small concentration from the domain where the concentration is on the order of U_0.

2. Finite initial conditions. Consider

$$u(x,0) \;=\; U_0 \text{ for } |x| \le \Delta \quad \text{and} \quad u(x,0) \;=\; 0 \text{ for } |x| > \Delta. \tag{9.30}$$

The corresponding solution of Eq. (9.21) has the following form:

$$u(x,t) = \frac{U_0}{2}\left[\operatorname{erf}\left(\frac{x+\Delta}{\sqrt{4Dt}}\right) - \operatorname{erf}\left(\frac{x-\Delta}{\sqrt{4Dt}}\right)\right]. \tag{9.31}$$

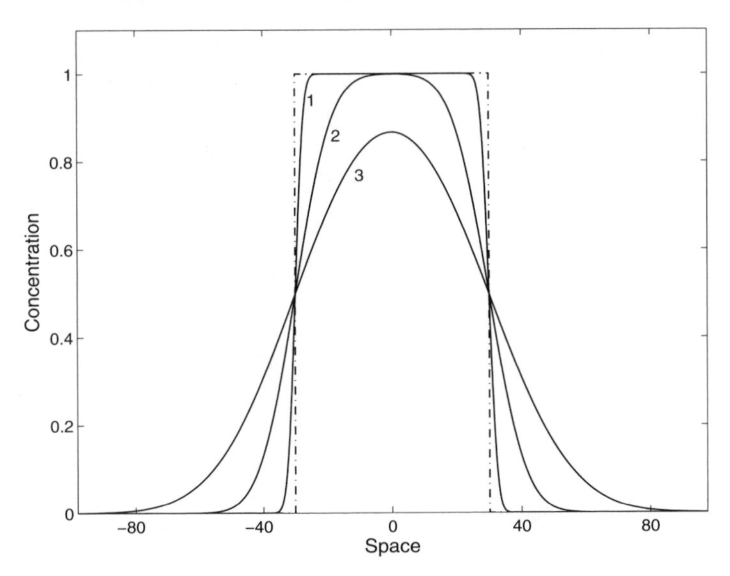

FIGURE 9.4: Solution (9.31) of the diffusion equation shown for $t = 2$ (curve 1), $t = 40$ (curve 2) and $t = 200$ (curve 3) for $D = U_0 = 1$, $\Delta = 30$; dashed-and-dotted line shows the initial conditions.

Solution (9.31) is shown in Fig. 9.4. It is straightforward to see that its properties appear to be a combination of those for the fundamental solution and for (9.28). In case Δ is small, the properties of (9.31) are close to (9.25). In case Δ is sufficiently large, at an early stage of the system dynamics, solution (9.31) describes diffusion of the left-hand and right-hand boundaries in the manner shown in Fig. 9.3. For larger time, it becomes similar to (9.25).

Note that the large-distance asymptotics of the solution (9.31) appears to coincide with the fundamental solution (9.25). Indeed, considering $\Delta/x \ll 1$, we observe that (9.31) contains a small parameter:

$$\frac{x \pm \Delta}{\sqrt{4Dt}} = \frac{x}{\sqrt{4Dt}} \left(1 \pm \frac{\Delta}{x}\right). \tag{9.32}$$

Correspondingly, applying the Taylor series expansion and taking into account (9.29), we obtain:

$$\mathrm{erf}\left(\frac{x \pm \Delta}{\sqrt{4Dt}}\right) = \mathrm{erf}\left(\frac{x}{\sqrt{4Dt}}\right) \pm \frac{2}{\sqrt{\pi}} \exp\left(\frac{x^2}{4Dt}\right) \cdot \frac{x}{\sqrt{4Dt}} \cdot \frac{\Delta}{x} + o\left(\frac{\Delta}{x}\right)$$

$$\approx \mathrm{erf}\left(\frac{x}{\sqrt{4Dt}}\right) \pm \frac{\Delta}{\sqrt{\pi Dt}} \exp\left(\frac{x^2}{4Dt}\right), \tag{9.33}$$

where $o(z)$ is the usual notation for the higher orders with respect to z. Substituting (9.33) to (9.31), we arrive at (9.25) where now $x_0 = 0$ and $G = U_0\Delta$.

In conclusion, we want to mention that, for all three solutions (9.25), (9.28) and (9.31), for any $t > 0$ the concentration appears to be positive at any position in space. That means that small disturbances propagate with an infinite speed. This is an essential artifact of the linear diffusion equation. A finite speed can arise when the diffusion coefficient is density-dependent, cf. Section 5.1.

References

Ablowitz, M.J. and A. Zeppetella (1979). Explicit solutions of Fisher's equations for a special wave speed. *Bull. Math. Biol.* **41**, 835-840.

Ablowitz, M.J., A. Ramani and H. Segur (1978). Nonlinear evolution equations and ordinary differential equations of Painlevé type. *Lett. Nuovo Cimento* **23**, 333-338.

Allee, W.C. (1938). *The Social Life of Animals.* Norton and Co, New York.

Allen, L. (2003). *An Introduction to Stochastic Processes with Applications to Biology.* Pearson Education, Upper Saddle River NJ.

Andow, D., P. Kareiva, S. Levin and A. Okubo (1990). Spread of invading organisms. *Landscape Ecology* **4**, 177-188.

Ardity, R., Y. Tyutyunov., A. Morgulis, V. Govorukhin and I. Senina (2001). Directed movement of predators and the emergence of density-dependence in predator-prey models. *Theor. Popul. Biol.* **59**, 207-221.

Aris, R. (1975). *The Mathematical Theory of Diffusion and Reaction in Pereable Catalyst, vols. 1 and 2.* Oxford University Press, Oxford.

Aronson, D.G. (1980). Density-dependent interaction systems. In *Dynamics and Modeling of Reactive Systems* (W.H. Steward, W. Harmon Ray and C.C. Conley, eds.). Academic Press, New York, pp. 161-176.

Aronson, D.G. and H.F. Weinberger (1975). Nonlinear diffusion in population genetics, combustion, and nerve pulse propagation. In *Lecture Notes in Mathematics* **446** (J.A. Goldstein, ed.). Springer, Berlin, pp. 5-49.

Aronson, D.G. and H.F. Weinberger (1978). Multidimensional nonlinear diffusion arising in population genetics. *Advances in Math.* **30**, 33-76.

Barenblatt G.I. (1996) *Scaling, Self-Similarity, and Intermediate Asymptotics.* Cambridge University Press, Cambridge, UK.

Barenblatt, G.I. and Y.B. Zeldovich (1971). Intermediate asymptotics in mathematical physics. *Russian Math. Surveys* **26**, 45-61.

Barenblatt, G.I., M.E. Vinogradov and S.V. Petrovskii (1995). Impact waves in spatially inhomogeneous open-sea ecosystems: localization and pattern formation. *Oceanology* **35**, 202-207.

Barenblatt, G.I., M. Bertsch, R. Dal Passo, V.M. Prostokishin and M. Ughi (1993). A mathematical model of turbulent heat and mass transfer in stably stratified shear flow. *J. Fluid Mech.* **253**, 341-358.

Bebernes, J. and D. Eberly (1989). *Mathematical Problems From Combustion Theory* (Appl. Math. Sci., vol. 83). Springer-Verlag, New York.

Benguria, R.D. and M.C. Depassier (1994). Exact fronts for the nonlinear diffusion equation with quintic nonlinearities. *Phys. Rev. E* **50**, 3701-3704.

Berezovskaya, F.S. and R.G. Khlebopros (1996). The role of migrations in forest insect dynamnics. In *Studies in Mathematical Biology – in Memorial of Alexander D. Bazykin* (E.E. Shnol, ed.). Pushchino Scientific Center, Pushchino, pp. 61-69. [in Russian]

Berezovskaya, F.S. and G.P. Karev (1999). Bifurcations of traveling waves in population taxis models. *Physics–Uspekhi* **169**, 1011-1024.

Braatne, J.H. and L.C. Bliss (1999). Comparative physiological ecology of Lupines colonizing early successional habitats on Mount St. Helens. *Ecology* **80**, 891-907.

Bressloff, P.C. and G. Rowlands (1997). Exact traveling wave solutions of an "integrable" discrete reaction-diffusion equation. *Physica D* **106**, 255-269.

Britton, N.F. (1986). *Reaction–Diffusion Equations and Their Applications to Biology.* Academic Press, London.

Burgers, J. (1948). *A Mathematical Model Illustrating the Theory of Turbulence.* Academic Press, New York.

Calogero, F. and J. Xiaoda (1991). C-integrable nonlinear partial differential equations. *J. Math. Phys.* **32**, 875-887.

Cannas, S.A., D.E. Marco and S.A. Paez (2003). Modelling biological invasions: species traits, species interactions, and habitat heterogeneity. *Math. Biosci.* **183**, 93-110.

Clark J.S., C. Fastie, G. Hurtt, S.T. Jackson, C. Johnson, G.A. King, M. Lewis, J. Lynch, S. Pacala, C. Prentice, E.W. Schupp, T.Webb III and P. Wyckoff (1998). Reid's paradox of rapid plant migration. *BioScience* **48**, 13-24.

Cole, J.D. (1951). On a quasi-linear parabolic equation occurring in aerodynamics. *Quarterly of Applied Mathematics* **9**(3), 225-236.

Courchamp, F., T. Clutton-Brock and B. Grenfell (1999). Inverse density dependence and the Allee effect. *Trends Ecol. Evolut.* **14**, 405-410.

Danilov, V.G. and P.Y. Subochev (1991). Exact single-phase and two-phase solutions for semilinear parabolic equations. *Teor. Mat. Fiz.* **89**, 25-47.

del Moral, R. and D.M. Woods (1993). Early primary succession on a barren volcanic plain at Mount St. Helens, Washington. *American Journal of Botany* **80**, 981-991.

Dennis, B. (1989). Allee effects: population growth, critical density, and the chance of extinction. *Nat. Res. Model.* **3**, 481-538.

Djumagazieva, S.K. (1983). Numerical integration of a certain partial differential equation. *J. Numer. Math. Phys.* **23**, 839-847.

Drake, J.A., H.A. Mooney, F. di Castri, R.H. Groves, F.J. Kruger, M. Rejmanek and M. Williamson, eds. (1989). *Biological Invasions: A Global Perspective.* John Wiley, Chichester.

Dunbar, S.R. (1983). Traveling wave solutions of diffusive Lotka–Volterra equations. *J. Math. Biol.* **17**, 11-32.

Dunbar S.R. (1984) Traveling wave solutions of diffusive Lotka–Volterra equations: a heteroclinic connection in $\mathbf{R}^4$. *Trans. Amer. Math. Soc.* **268**, 557-594.

Edwards, A.M. and A. Yool (2000). The role of higher predation in plankton population models. *J. Plankton Res.* **22**, 1085-1112.

Elton C.S. (1958) *The Ecology of Invasions by Animals and Plants.* Methuen and Company, London.

Fagan, W.F. and J.G. Bishop (2000). Trophic interactions during primary succession: herbivores slow a plant reinvasion at Mount St. Helens. *Amer. Nat.* **155**, 238-251.

Fagan, W.F., M.A. Lewis, M.G. Neubert and P. van den Driessche (2002). Invasion theory and biological control. *Ecology Letters* **5**, 148-157.

Fath, G. (1998) Propagation failure of traveling waves in a discrete bistable medium. *Physica D* **116**, 176-190.

Feller, W. (1971). *Probability and its Applications, Vol. 2.* Wiley, New York.

Feltham D.L. and M.A.J. Chaplain (2000). Traveling waves in a model of species migration. *Applied Mathematics Letters* **13**, 67-73.

Fisher, R. (1937). The wave of advance of advantageous genes. *Ann. Eugenics* **7**, 255-369.

Frantzen, J. and F. van den Bosch (2000). Spread of organisms: can traveling and dispersive waves be distinguished? *Basic and Applied Ecology* **1**, 83-91.

Gilpin, M.E. (1972). Enriched predator-prey systems: theoretical stability. *Science* **177**, 902-904.

Grindrod, P. (1996). *The Theory and Applications of Reaction–Diffusion Equations, Patterns and Waves* (2nd edition). Clarendon Press, Oxford.

Gurney W.S.C and R.M. Nisbet (1975). The regulation of inhomogeneous population. *J. Theor. Biol.* **52**, 441-457.

Gurney W.S.C and R.M. Nisbet (1976). A note on nonlinear population transport. *J. Theor. Biol.* **56**, 249-251.

Hadeler, K.P. and F. Rothe (1975). Traveling fronts in nonlinear diffusion equations. *J. Math. Biol.* **2**, 251-263.

Haken, H. (1983). *Advanced Synergetics.* Springer, Berlin.

Halvorson, J.J., E.H. Franz, J.L. Smith and R.A. Black (1992). Nitrogenase activity, nitrogen fixation, and nitrogen inputs by lupines at Mount St. Helens. *Ecology* **73**, 87-98.

Hengeveld, R. (1989). *Dynamics of Biological Invasions.* Chapman and Hall, London.

Herrera, J.J.E., A. Minzoni and R. Ondarza (1992). Reaction-diffusion equations in one dimension: particular solutions and relaxation. *Physica D* **57**, 249-266.

Higgins, S.I. and D.M. Richardson (1996). A review of models of alien plant spread. *Ecological Modelling* **87**, 249-265.

Higgins, S.I., D.M. Richardson and R.M. Cowling (1996). Modelling invasive plant spread: the role of plant-environment interactions and model structure. *Ecology* **77**, 2043-2055.

Höllig, K. (1983). Existence of infinitely many solutions for a forward backward heat equation. *Trans. Amer. Math. Soc.* **278**, 299-316.

Holmes, E.E. (1993). Are diffusion models too simple? A comparison with telegraph models of invasion. *Amer. Nat.* **142**, 779-795.

Holmes, E.E., M.A. Lewis, J.E. Banks and R.R. Veit (1994). Partial differential equations in ecology: spatial interactions and population dynamics. *Ecology* **75**, 17-29.

Horsthemke, W. and R. Lefever (1984). *Noise-induced Transitions. Theory and Applications in Physics, Chemistry and Biology.* Springer, Berlin.

Hopf, E. (1950). The partial differential equation $u_t + uu_x = \mu u_{xx}$. *Comm. Pure Appl. Math.* **3**, 201-216.

Jones, D.S. and B.D. Sleeman (1983). *Differential Equations and Mathematical Biology.* George Allen and Unwin, London.

Kanel, Y.I. (1964). Stabilization of the solutions of the equations of combustion theory with finite initial functions. *Matem. Sbornik* **107**, 398-413. [in Russian]

Kareiva, P. (1983). Local movement in herbivorous insects: applying a passive diffusion model to mark-recapture field experiments. *Oecologia* **57**, 322-327.

Kawahara, T. and M. Tanaka (1983). Interactions of traveling fronts: an exact solution of a nonlinear diffusion equation. *Phys. Lett. A* **97**, 311-314.

Kawasaki, K., A. Mochizuki, M. Matsushita, T. Umeda and N. Shigesada (1997). Modeling spatio-temporal patterns generated by *Bacillus subtilis*. *J. Theor. Biol.* **188**, 177-185.

Keener, J.P. (1987) Propagation and its failure in coupled systems of discrete excitable cells. *SIAM J. Appl. Math.* **47**, 556-572.

Keitt, T.H., M.A. Lewis and R.D. Holt (2001). Allee effects, invasion pinning, and species' borders. *Amer. Nat.* **157**, 203-216.

Kierstead, L. and L.B. Slobodkin (1953). The size of water masses containing plankton blooms. *Journal of Marine Reseach* **12**, 141-147.

Kinezaki, N., K. Kawasaki, F. Takasu and N. Shigesada (2003). Modeling biological invasions into periodically fragmented environments. *Theor. Popul. Biol.* **64**, 291-302.

Kolmogorov, A.N. (1931). Über die analytischen methoden in der wahrscheinlichkeitsrechnung. *Math. Ann.* **104**, 415-458. [English translation: On analytical methods in probability theory, in *Selected Works by A.N. Kolmogorov, vol. 2* (A.N. Shiryayev, ed.). Kluwer, Dordrecht, 1992, pp. 62-108]

Kolmogorov, A.N., I.G. Petrovsky and N.S. Piskunov (1937). Investigation of the equation of diffusion combined with increasing of the substance and its application to a biology problem. *Bull. Moscow State Univ. Ser. A: Math. and Mech.* **1**(6), 1-25.

Kot, M. (2001). *Elements of Mathematical Ecology.* Cambridge Univeristy Press, Cambridge, UK.

Kot, M., M.A. Lewis and P. van der Driessche (1996). Dispersal data and the spread of invading organisms. *Ecology* **77**, 2027-2042.

Larson, D. (1978). Transient bounds and time-asymptotic behaviour of solutions to nonlinear equations of Fisher type. *SIAM J. Appl. Math.* **34**, 93-103.

Lefschetz, S. (1963). *Differential Equations: Geometric Theory.* Interscience, New York.

Lewis, M.A. and P. Kareiva (1993). Allee dynamics and the spread of invading organisms. *Theor. Popul. Biol.* **43**, 141-158.

Liebhold, A.M., J.A. Halverson and G.A. Elmes (1992). Gypsy moth invasion in North America: a quantitative analysis. *Journal of Biogeography* **19**, 513-520.

Lotka, A.J. (1925). *Elements of Physical Biology.* Williams and Wilkins, Baltimore.

Lubina, J.A. and S.A. Levin (1988). The spread of a reinvading species: range expansion in the California sea otter. *Amer. Nat.* **131**, 526-543.

Malchow, H., S.V. Petrovskii and A.B. Medvinsky (2002). Numerical study of plankton-fish dynamics in a spatially structured and noisy environment. *Ecol. Modelling* **149**, 247-255.

Malchow, H., F. Hilker and S.V. Petrovskii (2004). Noise and productivity dependence of spatiotemporal pattern formation in a prey-predator system. *Discrete and Continuous Dynamical Systems* **4**, 707-713.

May, R.M. (1972). Limit cycles in predator-prey communities. *Science* **177**, 900-902.

McKean, H.P. (1975). Application of Browninan motion to the equation of Kolmogorov–Petrovskii–Piskunov. *Comm. Pure Appl. Math.* **28**, 323-331.

Mikhailov, A.S. (1990). *Foundations of Synergetics I.* Springer, Berlin.

Mimura, M., H. Sakaguchi and M. Matsushita (2000) Reaction-diffusion modelling of bacterial colony patterns. *Physica A* **282**, 283-303.

Mundinger, P.C. and S. Hope (1982). Expansion of the winter range of the House Finch: 1947-79. *American Birds* **36**, 347-353.

Murray, J.D. (1989). *Mathematical Biology.* Springer, Berlin.

Neubert, M.G. and H. Caswell (2000). Demography and dispersal: calculation and sensitivity analysis of invasion speed for structured populations. *Ecology* **81**, 1613-1628.

Newell, A.C., M. Tabor and Y.B. Zeng (1987). A unified approach to Painlevé expansions. *Physica D* **29**, 1-68.

Newman, W.I. (1980). Some exact solutions to a nonlinear diffusion problem in population genetics and combustion. *J. Theor. Biol.* **85**, 325-334.

Ognev, M.V., S.V. Petrovskii and V.M. Prostokishin (1995). Dynamics of formation of a switching wave in a dissipative bistable medium. *Technical Physics* **40**, 521-524.

Okubo, A. (1980). *Diffusion and Ecological Problems: Mathematical Models*. Springer, Berlin.

Okubo, A. (1986). Dynamical aspects of animal grouping: swarms, schools, flocks, and herds. *Adv. Biophys.* **22**, 1-94.

Okubo, A. (1988). Diffusion-type models for avian range expansion. In: *Acta XIX Congress Internationalis Ornithologici* (H. Quellet, ed.). University of Ottawa Press, Ottawa, pp. 1038-1049.

Okubo, A. and S. Levin (1989). A theoretical framework for data analysis of wind dispersal of seeds and pollen. *Ecology* **70**, 329-338.

Okubo, A. and S. Levin (2001). *Diffusion and Ecological Problems: Modern Perspectives*. Springer, Berlin.

Okubo, A., P.K. Maini, M.H. Williamson and J.D. Murray (1989). On the spatial spread of the grey squirrel in Britain. *Proceedings of Royal Society of London B* **238**, 113-125.

Otwinowski, M., R. Paul and W.G. Laidlaw (1988). Exact travelling wave solutions of a class of nonlinear diffusion equations by reduction to a quadrature. *Phys. Lett. A* **128**, 483-487.

Owen, M.R. and M.A. Lewis (2001). How predation can slow, stop or reverse a prey invasion. *Bull. Math. Biol.* **63**, 655-684.

Petrovskii, S.V. (1997). Localization of a nonlinear switching wave in an active medium with an isolated inhomogeneity. *Technical Physics* **42**, 866-871.

Petrovskii, S.V. (1998). Modeling of open-sea ecological impact: impact wave localization and pattern formation. *Environment Modeling and Assessment* **3**, 127-133.

Petrovskii, S.V. (1999a). Exact solutions of the forced Burgers equation. *Technical Physics* **44**, 878-881.

Petrovskii, S.V. (1999b) Plankton front waves accelerated by marine turbulence. *J. Marine Sys.* **21**, 179-188.

Petrovskii, S.V. and B.-L. Li (2003). An exactly solvable model of population dynamics with density-dependent migrations and the Allee effect. *Math. Biosci.* **186**, 79-91.

Petrovskii, S.V. and H. Malchow, eds. (2005). *Biological Invasions in a Mathematical Perspective* (Proceedings of the special session on biological invasions modeling held at CMPD, Trento, Italy, June 21-25, 2004). A special issue of *Biological Invasions*, in press.

Petrovskii, S.V. and N. Shigesada (2001). Some exact solutions of a generalized Fisher equation related to the problem of biological invasion. *Math. Biosci.* **172**, 73-94.

Petrovskii, S.V., A.Y. Morozov and E. Venturino (2002). Allee effect makes possible patchy invasion in a predator-prey system. *Ecology Letters* **5**, 345-352.

Petrovskii, S.V., H. Malchow and B.-L. Li (2005a). An exact solution of a diffusive predator-prey system. *Proc. R. Soc. Lond. A* **461**, 1029-1053.

Petrovskii, S.V., A.Y. Morozov and B.-L. Li (2005b). Regimes of biological invasion in a predator-prey system with the Allee effect. *Bull. Math. Biol.* **67**, 637-661.

Posmentier, E.S. (1977). The generation of salinity fine structures by vertical diffusion. *J. Phys. Oceanogr.* **7**, 298-300.

Protter, M.H. and H.F. Weinberger (1984). *Maximum Principles in Differential Equations.* Springer-Verlag, Berlin.

Rothe, F. (1978). Convergence to traveling fronts in semilinear parabolic equations. *Proc. R. Soc. Edin. A* **80**, 213-234.

Sachdev, P.L. (1987). *Nonlinear Diffusive Waves.* Cambridge University Press, New York.

Sakai, A.K., F.W. Allendorf, J.S. Holt, D.M. Lodge, J. Molofsky, K.A. With, S. Baughman, R.J. Cabin, J.E. Cohen, N.C. Ellstrand, D.E. McCauley, P. O'Neil, I.M. Parker, J.N. Thompson and S.G. Weller (2001). The population biology of invasive species. *Annu. Rev. Ecol. Syst.* **32**, 305-332.

Samarskii, A.A., V.A. Galaktionov, S.P. Kurdyumov and A.P. Mikhailov (1987). *Peaking Modes in Problems for Quasilinear Parabolic Equations.* Nauka, Moscow. [in Russian]

Sánchez-Garduño, F. and P.K. Maini (1994). Existence and uniqueness of a sharp front traveling wave in degenerate nonlinear diffusion Fisher–KPP equations. *J. Math. Biol.* **33**, 163-192.

Sander, L.M., C.P. Warren, I.M. Sokolov, C. Simon and J. Koopman (2002). Percolation on heterogeneous network as a model for epidemics. *Math. Biosci.* **180**, 293-305.

Scott, A.C. (1977). *Neurophysics.* Wiley, New York.

Sharov, A.A. and A.M. Liebhold (1998). Model of slowing the spread of gypsy moth (Lepidoptera: Lymantridae) with a barrier zone. *Ecological Applications* **8**, 1170-1179.

Sherratt, J.A. (1994). Chemotaxis and chemokinesis in eukariotic cells: the Keller–Segel equations as an approximation to a detailed model. *Bull. Math. Biol.* **56**, 129-146.

Sherratt, J.A. and B.P. Marchant (1996). Nonsharp traveling wave fronts in the Fisher equation with degenerate nonlinear diffusion. *Applied Mathematics Letters* **9**(5), 33-38.

Sherratt, J.A., M.A. Lewis and A.C. Fowler. (1995) Ecological chaos in the wake of invasion. *Proc. US Natl. Acad. Sci.* **92**, 2524-2528.

Sherratt, J.A., B.T. Eagan and M.A. Lewis (1997). Oscillations and chaos behind predator-prey invasion: mathematical artifact or ecological reality? *Phil. Trans. R. Soc. Lond. B* **352**, 21-38.

Shigesada, N., K. Kawasaki and E. Teramoto (1986). Traveling periodic waves in heterogeneous environments. *Theor. Popul. Biol.* **30**, 143-160.

Shigesada, N., K. Kawasaki and Y. Takeda (1995). Modelling stratified diffusion in biological invasions. *Amer. Nat.* **146**, 229-251.

Shigesada, N. and K. Kawasaki (1997). *Biological Invasions: Theory and Practice.* Oxford University Press, Oxford.

Skellam, J.G. (1951). Random dispersal in theoretical populations. *Biometrika* **38**, 196-218.

Stauffer, D. and A. Aharony (1992). *Introduction to Percolation Theory.* Taylor and Francis, London.

Steele, J.H. and E.W. Henderson (1992a). The role of predation in plankton model. *J. Plankton Res.* **14**, 157-172.

Steele, J.H. and E.W. Henderson (1992b). A simple model for plankton patchiness. *J. Plankton Res.* **14**, 1397-1403.

Strier, D.E., D.H. Zanette and H.S. Wio. (1996). Wave fronts in a bistable reaction-diffusion system with density-dependent diffusivity. *Physica A* **226**, 310-323.

Torchin, M.E., K.D. Lafferty, A.P. Dobson, V.J. McKenzie and A.M. Kuris (2003). Introduced species and their missing parasites. *Nature* **421**, 628-630.

Turchin P. (1998). *Quantitative Analysis of Movement.* Sinauer, Sanderland.

United States Bureau of Entomology and Plant Quarantine (1941). Japanese beetle. *Insect Pest Surv. Bull. U.S.* **21**, 801-802.

Vinogradov, M.E., E.A. Shushkina, E.I. Musaeva and P.Y. Sorokin (1989). A new invader to the Black Sea – ctenophore Mnemiopsis leidyi (A. Agassiz) (Ctenophora Lobata). *Oceanology* **29**, 293-299.

Volpert, A.I. and S.I. Khudyaev (1985). *Analysis in Classes of Discontinuous Functions and Equations of Mathematical Physics.* Nijoff, Dordrecht.

Volpert, A.I., V.A. Volpert and V.A. Volpert. (1994). *Traveling Wave Solutions of Parabolic Systems.* American Mathematical Society, Providence.

Volterra, V. (1926). Fluctuations in the abundance of a species considered mathematically. Nature *118*, 558-560.

Wang, M.-E., M. Kot and M.G. Neubert (2002). Integrodifference equations, Allee effects, and invasions. *J. Math. Biol.* **44**, 150-168.

Weinberger, H.F. (1982). Long-time behavior of a class of biological models. *SIAM J. Math. Anal.* **13**, 353-396.

Weiss, J., M. Tabor and G. Carnevale (1983). The Painlevé property for partial differential equations. *J. Math. Phys.* **24**, 522-526.

Wilhelmsson, H. (1988a). Simultaneous diffusion and reaction processes in plasma dynamics. *Phys. Rev. A* **38**, 1482-1489.

Wilhelmsson, H. (1988b). Solution of reaction-diffusion equations describing dynamic evolution towards explosive localized states. *Phys. Rev. A* **38**, 2667-2670.

Zeldovich, Y.B. (1948). Theory of propagation of flame. *Zhurnal Fizicheskoi Khimii* **22**, 27-48. [in Russian]

Zeldovich, Y.B. and G.I. Barenblatt (1959). Theory of flame propagation. *Combustion and Flame* **3**, 61-74.

Zeldovich, Y.B., G.I. Barenblatt, V.B. Librovich and G.M. Makhviladze (1980). *Mathematical Theory of Combustion and Flame.* Consultants Bureau, New York.

Index